Automation for Productivity

Automation for Productivity

Hugh D. Luke

Chairman and Chief Executive Officer
Reliance Electric Company

A Becker & Hayes Publication

JOHN WILEY & SONS
New York   London   Sydney   Toronto

Copyright © 1972 by Becker and Hayes Inc.
a subsidiary of John Wiley & Sons, Inc.

Library of Congress Cataloging in Publication Data

Luke, Hugh D
  Automation for productivity.
  Includes bibliographical references.
  1. Automation. I. Title.
T59.5.L84   658.5   72-5441
ISBN 0-471-55400-6

Printed in the United States of America

10 9 8 7 6 5 4 3 2 1

Preface

Although automation has long been employed to provide a multitude of reasonably priced products for the American consumer, more and more it becomes evident that its real character as a manufacturing systems approach needs to be examined carefully and broadly as an economic phenomenon. In this book the purpose is to examine automation technology in its broadest sense and develop not only an appreciation for it as such but also present the organizational "how-to" by which business management can more effectively utilize automation to improve productivity and combat rising costs in the years ahead.

Fundamentally, this book is addressed to industrial managers so as to provide top management approach for attaining success in automating. In addition, it highlights the manufacturing research and long-range planning that will be required for creating the new manufacturing technology so necessary for assuring success in future automation efforts.

One of the important facts emphasized in this text is that automation is *not* merely another kind or type of machinery. To effect true productivity improvement requires a fresh look at the *entire* production process or facility — as a completely integrated system. With the developments of the past few years, rapid advances in the technology and the "tools of automation" have brought this imperative goal within the reasonable grasp of manufacturing management in almost every segment of industry. However, to utilize this progress, it is necessary to cope with some of the new problems which have arisen as a result. These must be met and solved before the full value of automated operations can be realized.

The key to management success with automation in the past has often been found to be elusive. Present revelations indicate this is still true. In this book, based on extensive experience, the most important facets of the industrial automation policy needed are placed in proper perspective. With a fundamental understanding, the chief executive can examine corporate goals, objectives and philosophy in terms of the present and the future. He alone can set the course for the future for he must answer the critical question, "On the day we open, is our new plant *already obsolete?*"

This text emphasizes a practical approach to the job of tying down workable solutions to the problems which must be solved. Drawing on the broad experience and practical know-how our company has gained from working with many industries, it can help increase understanding of the principles of automation, the engineering challenges it presents — and the sizeable rewards which automation can bring.

Most significant, this book should make it possible for business and industry to focus a keener insight on the social and economic responsibilities implicit in management today. Since automation will help provide more quality goods with fewer industrial workmen to provide the needs of society, more persons will be transferring into the services and a higher standard of living. As this takes place, it is imperative in the author's mind that businessmen consciously seek to promote the well-being of the people associated with the enterprise they represent and also accept a more intimate responsibility to the broad public it serves.

Automation technology helps support many personal interests and ambitions in a challenging context when its dynamic character is fully realized. This book attempts to offer this practical understanding in easily understood terms for a wide range of utility. Top managers and chief executives in business and industry will find it an important addition to their library of basic texts.

Hugh D. Luke

Contents

Automation Economics and Technology

Management Practices and Policies for Automation

Major Industrial Change Influencing Automation Planning

Automation Economics and Technology

1

The Economics of Automation

Over the past three hundred years a widely diverse array of developments and inventions have laid the groundwork for modern-day automation. In manufacturing, processing, handling, and information systems today, the basic ideas of automatic operation conceived by such men as Evans, Watt, Arkwright, Jacquard, Perkins, Francis, and many others have merged to create the basis for the technology of automation. It is a natural and satisfactory answer to many of the problems which have arisen in the past several decades with the increasing American ambition for a higher standard of living.

Automation in business and industry has developed largely in a climate of free competition and a growing demand. An additional factor has been the difficulty encountered in processing and manufacturing by hand methods within acceptable quantity, quality, and cost levels. Wherever suitable cost and output were impractical of attainment, automation has gradually developed.

1

WHAT IS IT?

The word "automation" is a contraction of the words automatic-operation. It implies the process of doing things automatically. It is not synonymous with any other word. It does not merely mean mass production; mass production is volume manufacture of interchangeable products. In the 1840s, Robbins, Kendall, and Lawrence of Windsor, Vermont, mass produced rifles on a truly interchangeable basis completely by hand methods.[1]

Automation is based upon but goes a long step beyond mere mechanization. Mechanization simply means doing things with or by machines, not *necessarily* automatically. True automation implies continuous or cyclic arrangement for manufacturing, processing, or performing services as automatically as is *economically practical* or necessary.[2] Hence, the premise of this book.

The primary feature of mass production is standardization of component characteristics so as to permit complete interchangeability. This means that any product or element in a manufactured lot can be substituted for any other at random. The technique of mass production permits production of piece-parts at diverse locations to specified tolerances so that all are sufficiently identical to permit random use or random assembly with other products. Its basic feature is the elimination of separate individual fitting up in manufacture and assembly. Mass production techniques constituted a major step in creating the possibility of providing large quantities of complex products with superior quality and uniformity at lower cost in time and money. But today, hand methods often fail to fulfill the requirements because of economics, market demand, speed, safety, and other factors. The solution is the succeeding step in the manufacturing picture – automation.

From the standpoint of the workman, automation eliminates the undesirable characteristics of mechanization in which the operator functions as an integral mechanical part of the production cycle. Instead it makes the operator a skilled director of an integrated production sequence; it requires greater knowledge of the product; it calls for increased responsibility; and returns in large measure pride in knowledge and workmanship.

Automation largely aims to take single separate processing operations and link them into an automatic continuous line. It can involve a few separate operations or all operations from raw materials to finished product. For example, instead of making pipe by a series of separate steps, steel billets enter the rolling mill, are rolled into sheet; the sheet is formed, butt welded, cut to length and threaded, to emerge as finished pipe used for plumbing. By eliminating separate handling operations the pipe is produced at speeds of more than 17 miles an hour.

There are many examples of automation on hand today. As we run through only those everyday necessities such as toothpicks, matches, paper, flour, breakfast cereals, beverages, food products, chemicals, hardware, and on down the list, the story is impressive. But, today the accomplishments in these areas are being transplanted into other production fields. Automation *to some degree* can be found almost everywhere, from producing shovel handles to warehousing products. And the results are always similar — better products and distribution at lower unit costs.

AUTOMATION IS AVAILABLE NOW

Automation is not a thing of the future. It is available now for solving many of our manufacturing and distribution cost problems. A brief listing of some of today's automation accomplishments provides a glimpse into the many present areas of use throughout industry.

There are consumer product plants such as one turning out appliances at the rate of one every 30 seconds with 2500 automatic and semiautomatic machines tied together with over 25 miles of conveyors.

Single paper mills, for instance, produce as much as 500 miles of facial tissue daily, six feet wide, on continuous automatic equipment.

By continuous automatic production in several steps from raw materials to finished parts, the lamp bulb industry produces an estimated 4 billion lamps per year.

Automated systems successfully recover high-purity iron from low-grade ores in our relatively depleted reserves. Almost totally automatic operation — with a per ton expenditure of electric power several thousand times greater than in open-pit mining — makes these processes economically competitive with foreign high-grade ores and improves the automation feasibility of subsequent steel production operations.

Telephone drop wire is produced on a round-the-clock basis at more than 2 billion feet a year. Copper is electroformed around steel base wire continuously, the wire is cleaned, lead and brass plated, inspected, and wound on reels at rates in excess of 3000 fpm. Extruded insulating cover is applied in a continuous automatically controlled operation.

Through the use of computers, paperwork in widespread groups of plants is closely coordinated. Sales data, office data, and shop orders come from the same master program, eliminating error and lost time. Orders can be placed and delivery dates set within several hours compared to the weeks or months previously commonplace.

Numerically coded tapes as well as computers are widely used to control various kinds of equipment. At present, machine tools, singly or in groups, follow instructions punched into a tape or provided by direct computer control. Machines use the digitized or computerized data to drill holes; machine surfaces;

select rivets from automatic feeders; place and drive the rivets under precisely controlled conditions; pick, bend, and connect electronic wiring; control punching operations; direct transfer machine functions; operate complex assembly machines; carry out performance and/or quality testing sequences; and control processing operations.

In warehousing, packages are carried into storage, delivered into specific areas, passed between floors, drawn out, and fed to delivery stations all under computer control of switches, relays, electric eyes, and other control devices. Work is progressing in computerized sorting and picking operations, normally a tedious and costly process which today sets a drastic limit on speed and cost in distribution of products.

In more than one plant today, production is an automatic computerized operation from raw materials to warehousing and shipping.

TYPES OF AUTOMATION SYSTEMS

In the achievement of automation, groups or sequences of processing operations, automatic mechanisms, or machines and control devices are brought into a single system to produce continuous or cyclic operation. Wherever two or more automatic machines are tied together with overriding automatic controls to create self-feeding, a self-initiating and self-checking progressive production process, an automated system is created. Material, data or pieces can be introduced into such a system manually or automatically and the processing steps carried out without manual intervention to completion. Generally, almost any process can be automated to some degree. Actually, it is possible to automate a single operation, a sequence of operations, a whole department or a plant.

Automation can be segregated into several types. It is possible to create an automatic batch system or an automatic continuous system. One chemical process, for instance, may be more economically carried out in batches while another is most economical when produced continuously. Process characteristics as well as production economics dictate the best method. Thus, job-lot type operations may require batching arrangement with amenity to continuous change while mass production operations may be more economically carried out on a continuous basis.

Secondly, automation systems may be set up with end-control or in-process control. With end-control, processing is completed before testing, checking, or gaging is done. This is suitable with many processes and feasible with others. However, where precision output is necessary, in-process control is desirable for economic reasons. Material being processed is under continuous control — metal parts, for instance, are gaged in the machine, and correction or size control is accomplished during the operation, thus production of scrap is prevented.

Chemical and similar processing operations are held under continuous sensing and measuring instrument control, and necessary corrections are fed back to the equipment continuously.

WHAT STIMULATES AUTOMATION?

It is fundamental that management keep in sharp focus the main reasons that can create the need to automate regardless of the scale of operations or their general character. By and large, it is a result of either the basic need to meet a mass market demand or to make possible a product that can be merchandized at an acceptable price, regardless of quantity.

Today, the only reason that certain candy makers are able to deliver 5-cent candy bars made with real chocolate is automation. Fully automatic production lines permit economic operation in spite of the inflated cost of ingredients.

As cost of raw materials rises along with wages, only one end result can create an increased living standard. Commodity prices must hold the line or rise at a lesser rate than wages to create the desired differential between real income and cost of living. To make this possible, production and distribution costs must be cut. All production operations which do not help in creating product value must be eliminated or minimized as to time/cost — such operations include handling, inspecting, packaging, and warehousing.

In seeking to create higher quality products for lowest possible unit costs, innovative automation offers rather dramatic returns. The important achievement lies in the increased production gained from lower scrap losses and lost time, better and more uniform products, improved machine use, more economic use of materials, and simplified steps in reaching the customer.

While it is impossible to do everything automatically, there are quite a few areas where it is not only possible but desirable. Machines met the need when slave labor was eliminated and today automatic machines can eliminate monotonous and useless labor. As living standards rise and population grows, the mass market demand creates a growing pressure for quality goods. Outmoded hand methods cannot afford the necessary quantities, prices, or availability. Under a free competitive system, this trend eventually leads to automation.

Modern bakeries provide an interesting example. Conventional mass bread baking involves considerable wastage in stale loaves. Up-to-date installations make bread production completely automatic to the point where the bread reaches the customer absolutely fresh. Loaves are produced, baked in ovens, packaged, quick-frozen, and passed into storage continuously. Additional cost savings are achieved through the possibility of making longer production runs before changing to another type of loaf.

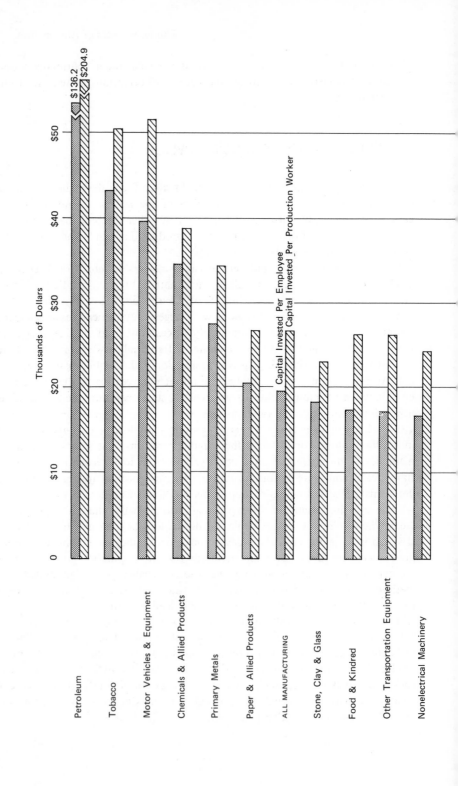

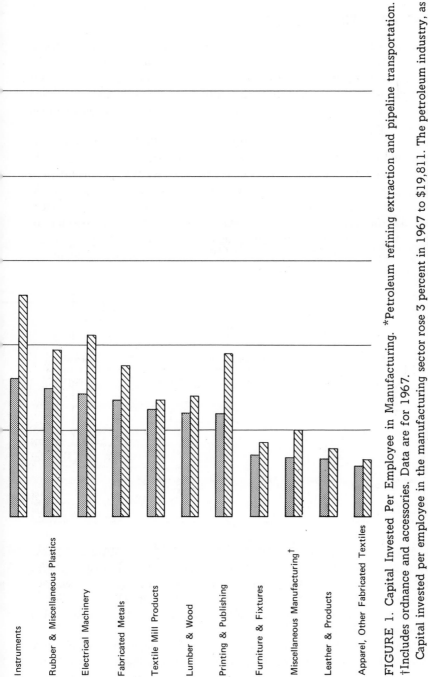

FIGURE 1. Capital Invested Per Employee in Manufacturing. *Petroleum refining extraction and pipeline transportation. †Includes ordnance and accessories. Data are for 1967.

Capital invested per employee in the manufacturing sector rose 3 percent in 1967 to $19,811. The petroleum industry, as in the past, had the highest capital invested per employee, $136,178; this was an increase of 13 percent over 1966. Investment per employee showed the largest annual increase in the "other transportation equipment" group, up 20 percent, reflecting a 30 percent increase in capital but only a 7 percent rise in employment. The largest decline in 1967 occurred in the tobacco industry, down 22 percent, resulting from reductions in capital accompanied by increases in employment.

AUTOMATION INVESTMENT

In order to achieve this productivity advance with automation, a great change has taken place in cost of the tools of production. In the oil industry, for instance, a typical plant invests over $200,000 in tools for each worker — over 900 percent more than at the turn of the century, see Figure 1. And instead of a 60-hour week there is the less than 40-hour week with hourly pay greater than former daily pay.

Along with such investment in tooling has come radically increased electric power to aid men's efforts. In average manufacturing operations, a typical 1944 plant used about 4½ h.p. of driven machinery per man. By 1953 this had risen to 9½ h.p., now to 25 h.p. But, power for automated operation has sparked a much faster rise. Today, automated plants using over 100 h.p. per worker are commonplace and individual cases are on hand where one man controls 20,000 h.p. or more of integrated machinery. Pumps using electric motors up to 9000 h.p. are no longer unusual; sizes ranging up to 17,000 h.p. for centrifugal compressors are in everyday use.

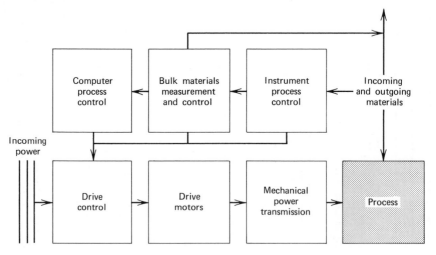

FIGURE 2.   The Automation Process.

Successful applications of this complexity require study of the manufacturing process in its entirety as a multi-stage development — a series of fundamental manufacturing operations carried on as an integrated system. This new attack on the problem, as shown in Figure 2, contrasts sharply with the departmental or piecemeal approach common in the past.

In developing and implementing this approach it is imperative to recognize that the total system consists of three major subsystems that may be inextricably interwoven into the final whole. The three subsystems are: processing or making, handling, and controlling, as shown in Figure 3. Any one or all of these functions may be critical to success, Figure 4.

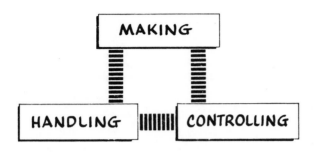

FIGURE 3.   Integrated Manufacturing.

For the study and development of the most profitable areas to automate it is desirable to recognize the possible need for reorganization in order to breach traditional internal departmental barriers. A manufacturing group, properly organized and backed by authority from the top, will be needed to carry out the necessary internal activities.

Productivity and/or profit improvement via automation may never be realized unless there is a central philosophy to keep all efforts on target and in proper balanced relationship. A great part of the general accomplishment in controlling manufacturing quality and costs is influenced directly by the equipment policy that prevails. Although it often appears remote, a major objective is to maintain an acceptable profit margin in face of quality demands rising along with material and labor. To ascertain the most economic approach, all detailed cost elements must be evaluated. Accounting methods may require some study. Machine-hours rather than man-hours may be the most critical new factor. In-process inventory may be a key element. Total manufacturing costs, door to door, must be analyzed. The clue is "It *pays* to know your true costs." Most don't until it is too late!

The major question today then is: Do you *really* know where the automation payoff will be? The answer is you don't know and can't know unless a thorough feasibility study is made covering all phases of the manufacturing system to pinpoint the steps to take, where, and in what sequence — conceivably a master plan. Management planning must pinpoint the necessary steps to be taken and provide the means by which it can be carried out.

This includes recognition of the return on investment made possible by the contributions of new technology from manufacturing research and/or outside engineering expertise. Finally, it is in this area that the company must enforce a reasonable policy on acquisition of new capital equipment. There is often little

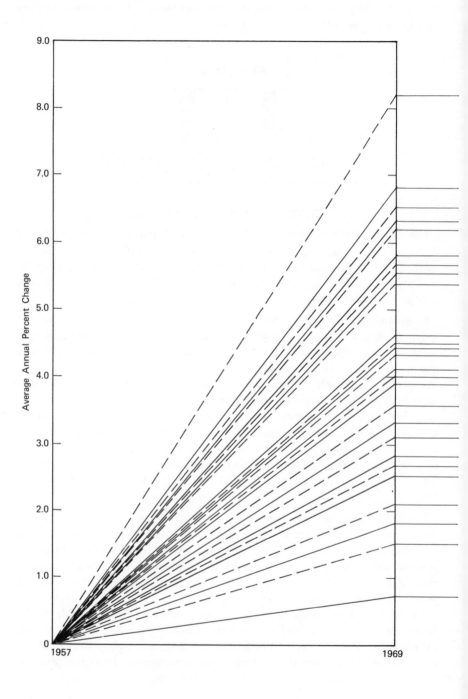

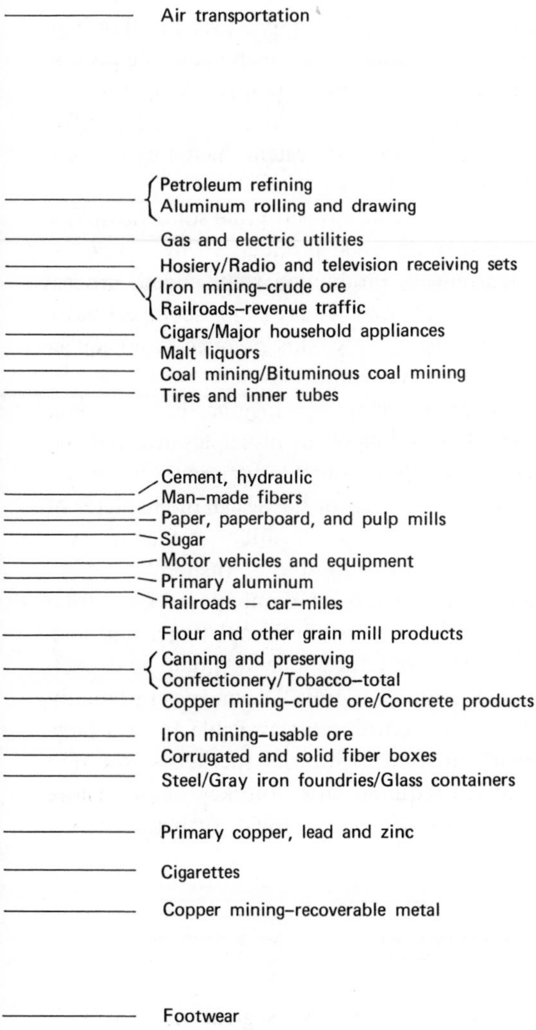

———————— Air transportation

{ Petroleum refining
{ Aluminum rolling and drawing

———————— Gas and electric utilities
———————— Hosiery/Radio and television receiving sets
———————— { Iron mining–crude ore
{ Railroads–revenue traffic
———————— Cigars/Major household appliances
———————— Malt liquors
———————— Coal mining/Bituminous coal mining
———————— Tires and inner tubes

———————— Cement, hydraulic
———————— Man–made fibers
———————— Paper, paperboard, and pulp mills
———————— Sugar
———————— Motor vehicles and equipment
———————— Primary aluminum
———————— Railroads — car–miles

———————— Flour and other grain mill products
———————— { Canning and preserving
{ Confectionery/Tobacco–total
———————— Copper mining–crude ore/Concrete products

———————— Iron mining–usable ore
———————— Corrugated and solid fiber boxes
———————— Steel/Gray iron foundries/Glass containers

———————— Primary copper, lead and zinc

———————— Cigarettes

———————— Copper mining–recoverable metal

———————— Footwear

FIGURE 4. Growth in output per man-hour in selected industries for 1957-69. (From Bulletin 1680, U. S. Department of Labor, Bureau of Labor Statistics, 1970.)

or no payback on equipment purchased solely on the basis of a low bid. Innovation – the heart and soul of automation – carries in it the ubiquitous seeds of profitable operation but is never revealed in a dollar price. *Best* answers can be had only from competent firms that can work cooperatively with the buyer's manufacturing group. Full knowledge of the problem is necessary to attain the expected profit potential, and this can show up *anywhere* in the total system.

Fundamental to the master plan is the fact that it provides for thorough "before-the-fact" planning and detailed execution of the manufacturing process stages and avoids costly "after-the-fact" correction of poorly planned or unplanned operations. The word "automation" implies just that – a broad, carefully engineered development for an overall processing system, including concern for product improvement with lower-cost processing.

Conventional approach to manufacturing has always permitted some flexibility and rather frequent correction of poorly designed products to overcome excessive costs. Machines could be rearranged, production steps added, process cycles readjusted, or personnel added, or removed. Today's more complex automated systems and lines have proved, conclusively, this practice is no longer either satisfactory or economical in most areas of industry.

Bringing the basic elements of automation technology together in the correct manner and balance requires a good understanding of the principles involved and a wide knowledge of materials and the methods suitable for processing them. The master plan procedure must include an analysis of the design or character of the product, the operating requirements, and the practicability of manufacture with known methods or new methods which are evolved to do the job.

Since no one company or single technical project group possesses a total knowledge of all the technology available for successfully completing an automation project, it is obvious that extensive use of outside talents should be expected. In reality, the automating company requires a project team to act primarily as a coordinating group to interface the contributions available from a large number of outside sources since few automation engineering specialists exist that are in a position to take on the total job required on a "turnkey" basis. Those that do provide this service must for economic reasons provide such service on a consulting or R&D basis.

TOP MANAGERS TAKE THE LEAD

As with any extremely complex system there is always great risk involved. Entering into the sophistication of automation technology necessitates thorough planning. It is *not* merely a matter of buying another piece of equipment. It is in reality an involvement with a total systems concept that spreads its tentacles into every area of the manufacturing operation.

Top executives and managers must genuinely understand and become involved. They must be dedicated to and support any such project resolutely. Only top management leadership can help surmount the obstacles and insure success. Getting results from any computerized system actually is not so much technical as it is managerial and organizational.

Maximizing the ability of current computer technology requires development of a total systems approach to effect a new total solution to the productivity goal in mind. In general, the fundamental philosophy must be that of any competitive business — profit or perish.

FOOTNOTES

1 *Precision Valley* by Wayne G. Broehl, Jr., Prentice-Hall, Inc., Englewood Cliffs, N.J., 1959, p. 5.

2 "Automation and the American Consumer" by Roger W. Bolz, *Automation*, April 1964, "What Automation Means to America," p. 16.

REFERENCES

Mechanization Takes Command by S. Giedion, Oxford University Press, 1948.

"Oliver Evans and His Inventions" by Coleman Sellers, Jr., *The Journal of the Franklin Institute,* Philadelphia, vol. 92 (1866).

American Science and Invention by Mitchell Wilson, Simon and Schuster, New York, 1954, p. 55.

Facts and Fancies about Automation by P. M. Boarman, NAM Industrial Relations Sourcebook Series, NAM, New York, 1965.

Technology and the American Economy, vol. 1, Report of the National Commission on Technology, Automation and Economic Progress, February 1966.

Automation by Jack Rogers, Institute of Industrial Relations, University of California, Berkeley, 1958.

Industry Case Examples

CASE 1A

High Speed Automated Mailroom System

Two hundred bundles a minute roll to truck-loading positions at the *Arizona Republic* and the *Phoenix Gazette* mailroom using an automated Bundle Escort System Totalizer (BEST) distribution method.

Need for greater speed and lower costs in the handling of bundles is being felt by small, medium, and large newspapers around the country. Growing circulations, more regional editions, and special inserts complicate the rapid handling of multiple products at one time.

This system has a maximum capacity of 200 bundles per minute. All twelve truck-loading positions at the Arizona dailies, plus one storage position, are available for assignment from any of the seven tying machines at any time, as shown in Figure 5. Five stackers feed counted stacks of newspapers to the tying machines. A sixth stacker and an eighth tying machine are to be added at a later date. All bundles are counted at each tying station and a count also is automatically maintained of the bundles sent to each truck, Figure 6.

The Sta-Hi system basically enables the distribution of a different product from each of many sources, i.e., tyer, storage, or inserter to the truck-loading area at any time. The system is a combination of synchronized conveyors controlled to escort predetermined quantities of bundles from mailroom tying lines to truck-loading positions, in Figure 7. The value of the system is its extreme flexibility in distributing bundles from any tying line in the system to any truck position at any time, without regard to how many tying lines are in simultaneous use or how many different products are being tied and distributed.

Routing of bundles to truck positions, or into storage, is automatic in response to preset bundle quantities. When the quantity to the truck being loaded is reached, the system will automatically route bundles to the next selected position.

There is built-in protection against obsolescence. Changing the locations of tying lines requires only that the system be resynchronized. The system can be modified to distribute around corners and in other than straight-line patterns.

All bundles entering the BEST system are counted and a total for each tying position is recorded on a master console, classified as the heart of the operation,

14

as shown in Figure 8. The console, governing assignments along a synchronized belt conveyer which moves at 334 fpm, activates pneumatic bundle deflectors positioned at each truck-loading point. Operation is based on the measurement of distance required for any bundle, or group of bundles, to go from each source or tying position. This is done by a combination of solid-state electronic equipment, a motor control center, and switches.

Part of the control system includes electronic memories to escort destination-change signals. This enables the last bundle to the current truck being loaded, and the first bundle to the next truck, to be deflected at their respective unloading points. Destination changes can be made without disrupting tying operations.

CREDITS: Sta-Hi Corporation, a Republic Corporation Company.

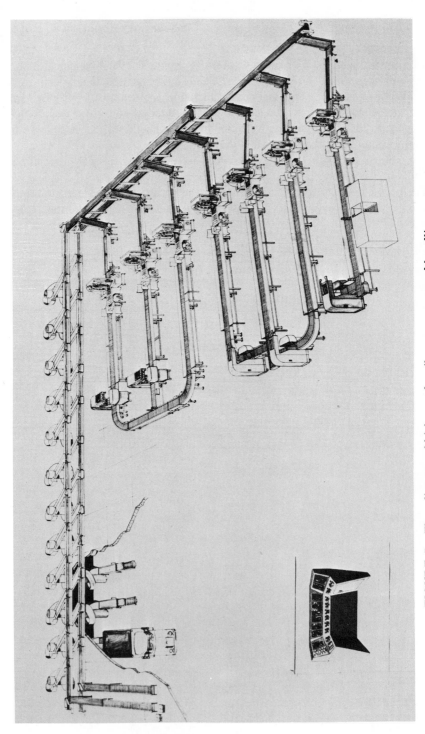

FIGURE 5. Flow diagram of high-speed mail room automated handling system. (Drawing from Sta-Hi Corporation, Newport Beach, California.)

FIGURE 6. Overall view of mailroom at Phoenix Republic and Gazette, showing all seven production lines in operation at one time. (Photograph from Sta-Hi Corporation, Newport Beach, California.)

FIGURE 7. Two-speed accumulator conveyor at right gathers tied bundles in trains of up to eight bundles, then ejects them on high-speed distribution belt. (Photograph from Sta-Hi Corporation, Newport Beach, California.)

FIGURE 8. Movement and count of all tied bundles to truck-loading positions is controlled from this centrally located console. (Photograph from Sta-Hi Corporation, Newport Beach, California.)

CASE 1B

The Automated Egg

Perhaps few manufactured products present the array of critical problems encountered in automating the production of farm products. The automated "egg ranch" is a good case in point.

Typical of many automated egg producers, Avian Bates, Inc., is the only "in-line fully automated egg farm" in the northeast part of the U.S.A. As with any automated manufacturing operation, lowest unit cost and highest productivity of egg production have been achieved through the ingenious system of equipments shown in the selection of views in Figures 9 - 14.

Avian Bates's revolutionary operation involves 280,000 laying hens. As the hen lays an egg, the egg drops onto a production belt, is carried through various service areas where it is cleaned, graded, and packaged. From laying to shipping, the egg is untouched by human hands.

CREDITS: Franklin National Bank and Big Dutchman, a Division of U.S. Industries, Inc.

FIGURE 9. Double-deck cages of birds.

FIGURE 10. Double-deck transveyors and control cross-belt.

FIGURE 11. Conveyors for volume egg collection control.

FIGURE 12.   In-line handling and central egg collecting.

FIGURE 13.   Egg diverter and orienter.

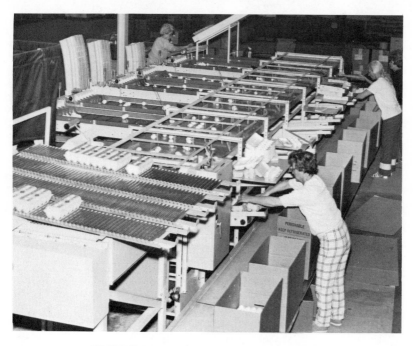

FIGURE 14.   Final collection and packaging.

2

Automation of Materials Processing

From the general purview of automation outlined in Chapter 1 it is rather obvious that automation as a totality is intimately involved with a production process of one kind or another. One or a series of basic materials processing operations are engineered into a continuous or automatic batch-type system.

From a cost standpoint it has been emphasized repeatedly over the years past that it may prove disastrous to automate the wrong process. The same is true with a poorly thought-out series of processing procedures. The maximum productivity can be realized where unnecessary processing steps can be eliminated or where even a larger number of different operations that offer better adaptability to continuous systems can be substituted, Figure 1.

It is possible, and the cases are far too numerous, to have brilliantly engineered production lines where the automation accomplishment has provided no improvement on overall out-the-door manufacturing costs. Here, the chances are good that the automation effort was concentrated on solving an interesting or

Production	Increases total quantity.
	Increases output per worker.
	Increases rate of production.
Labor	Decreases unskilled labor.
	Increases skilled labor.
	Improves worker conditions, relations, and attitudes.
	Increases safety.
Product	Decreases manual handling.
	Decreases unit costs.
	Increases quality.
	Increases uniformity.
Profits	Reduces costs.
	Increases net earnings.

FIGURE 1.   Results of successful automation.

challenging control problem at considerable expense without consideration of the important processing function.

SELECTING THE RIGHT PROCESSES

Despite much thought to the contrary, many processes are eminently suited *only* to manual operation. If the product is designed in such a manner that it must be made by manual processes, automation is usually inadvisable. To attempt the duplication of manual dexterity in a machine is almost always destined to end up in the loss column.

It must be realized that wherever handcraft methods are used there will be difficulty in meeting not only competition on a local or foreign basis but in meeting the demands for successful automation. A prominent maker of lenses has commented that "despite the fact that labor costs are by all means the most important single element in our company's cost structure, we are able to compete with manufacturers in foreign markets and expect to increase our share of these markets as well as our unit sales in the years immediately ahead. Through technological advances, our optical division has been able to export lenses, a fact of which its people are very proud. When you consider that the value of a lens is almost entirely in its workmanship, you can understand that pride. We once estimated that in a lens selling for $87, the value of the glass and metal was only 43 cents, the remaining cost representing labor. Without the spur of foreign competition, it is doubtful whether the techniques that have made exports possible would have been developed."

Two imperatives are present here: first, it is important to know, understand, and select or develop the best process or processes; second, it is equally important that product design, wherever feasible, be modified as much as possible to match the process capabilities. This latter area will be dealt with in a subsequent chapter.

This procedure of process selection can often be directly influenced by the conditions imposed by automation technology itself. To be suitable for operation with fast, accurate control -- including feedback and in some cases computer control — all production operations must be compatible and often may require a minimum cycle time. Extensive development of processing operations may need to be undertaken in order to insure compatibility in the overall sequence. Thus, in fact, the product's physical characteristics can become a direct function of the ability to *process* in the allowable time while meeting the set product specifications.

Perhaps the importance of knowledge over the widest possible range of useful processing methods is most dramatically emphasized by the stringent requirements made by process time cycles. Obviously, unless the available processes to be used are capable of continuous operation, the road to success with automation lies in the direction of integrating the desired series of separate stages into a closely controlled system. Here, time cycle of each stage of the integrated system is critical. Those who have either ignored this requirement or failed to recognize its critical character have experienced some nagging failures.

A good typical example of this prob'em is the fast cycling integrated system that included an adhesive bonding station. The bonding required a set period many times the output rate of the remainder of the machine operations. This one operation was improperly evaluated and led to total loss of the investment. To insure financial success *every* process step must meet system requirements as well as product functional specifications. Flexibility in the method used in meeting these specifications is absolutely essential.

From a study of this materials processing phase of automation, the evidence shows it is desirable to recognize from a financial standpoint that there are two major situations that emerge. These two situations in large part determine the amount or degree of risk involved:

1. Automated systems that can be created around processes that are well defined, well understood, and easily controllable. Unless the processes are well understood, it is typical to find oneself unexpectedly catapulted into a major process research program to save an entire automation plan.

2. Automated systems that involve one or many new, untried, or developmental processes. Such programs are definitely in the high risk area and should be recognized as research and development projects for which top management must accept the responsibility. Without access to a highly competent technical staff, which is in a position to call in competent outside help whenever and wherever needed, the project can easily become impossible.

THE ECONOMIC DECISION

Thus it can be seen that automation is not an end in itself. The much talked about "completely automatic factory" is not the standard by which a given process should be measured. It does no good to extensively automate an operation or process if such does not result in progress toward achieving the purpose of the enterprise. The purpose of any enterprise is profit. The goal of automation, therefore, must be to increase the productivity of an enterprise.

The various ways productivity can be maintained or increased in an enterprise are the basic reasons why automation techniques are employed. Automation is effective in attaining any, all, or any combination of the following objectives: (1) increase in production volume, (2) reduction of production costs, (3) increase in product quality, (4) elimination of process hazards and/or worker tedium, and (5) production of a unique product that could not be manufactured by conventional techniques.

Increase in Production Volume. Anticipating and/or responding to a marked increase in the market demand for a product — whether that product is a raw material, a tool for production, or a consumer item — might require more than intensifying existing efforts. It may be impossible to attain the production capacity necessary by adding more workers and/or extending conventional production lines. Even if these alterations would result in the volume required, product quality might suffer and/or costs might skyrocket. Often, new manufacturing methods are necessary — which is where automation might come in.

Reduction of Production Costs. Often the introduction or success of a product hinges on whether or not its market price is low enough. Or, shifting conditions in the marketplace (i.e., newer competing products, etc.) may necessitate a price reduction if sales are to be maintained. In any such case, production costs must be reduced. A change in technique becomes imperative.

By reducing the direct labor content in a process, automation techniques can reduce production costs quite effectively. Also, the application of automation methods can, in most instances, sharply reduce material costs. Continuously more accurate cutting and assembly operations, for example, will limit scrap material to a minimum. Another way automation can reduce costs is by a more accurate assessment and control of inventory. Automatic storage equipment, together with the complementary information systems, can severely cut the cost of maintaining and handling raw material or product inventories. In addition, by reducing setup times, economic runs can be reduced and thereby cut inventory costs.

Above and beyond these routes to savings is the fact that automation sharply increases the average and total output per worker. By making more powerful tools available to workers, automation techniques greatly multiply worker productivity. This accomplishment not only allows for reductions in prices — or for holding the line on prices where price increases would be necessary otherwise — it also enhances the profitability of the enterprise, which means higher worker wages and more capital for investment.

Increase in Product Quality. An increase in product quality may be necessary for many reasons. To hold and/or increase a product's share of the market, for example, usually requires a constantly improving product in the face of increasingly intense competition. Manual operations not only lack the *accuracy* of a mechanized procedure, but the *variations* in quality which are an inevitable part of human labor can be eliminated by the proper automatic controls. Numerically controlled machine tools, for example, can machine high quality parts to the same exact tolerances every time.

Inspection is always an expensive operation, which in many cases can be incorporated into machinery. Some automatic inspection operations can be accomplished at very high rates of speed with incredible accuracy, far beyond the capability of human beings. The more stringent standards and more rapid rates of automatic testing and inspection equipment result in a repetitive product quality that cannot be matched in any other way.

Elimination of Process Hazards and Worker Tedium. Some jobs, without automatic techniques, would not be done at all or would be accomplished with extreme human sacrifice. Other tasks, while not as absolute in their possible consequences, threaten the health and limbs of workers in such a way that automating results in worthwhile "people benefits" as well as tangible cost benefits.

The operator of an N/C machine tool, for example, oversees the cutting operation instead of participating in it. He not only turns out a better product, but he does so with no danger to himself. The testing of toxic chemicals and gases can be accomplished at a distance, automatically, without irritation or other detrimental effects to workers. By defusing dangerous operations, insurance rates could drop and employee morale would certainly rise.

Many production tasks, while not dangerous, are too repetitive and/or tedious to stimulate the required performance of employees. By automating such jobs — freeing workers for more interesting work — occupational malaise would be prevented, absenteeism would probably decline, and worker performance would improve.

Production of a Unique Product. Many existing products could not have been produced in the numbers or with the qualities necessary without automation. Production consists of, and results in, a steadily progressing hierarchy of more capable tools and better performing products. Each stage of production capability makes tools and products possible which would have been impossible otherwise. Computer technology is a perfect example of how automation has become indispensible. The minicircuits, memory elements, etc., that are a part of every computer could not have been produced in the numbers required without the exacting, high-volume capability of automatic machining, assembly, and testing and inspection techniques. In turn, computers constitute a new level of control capability which is resulting in a new level of productivity and profitability. With the computer, as with previous automation methods, processes will be controlled that could not have been controlled before and, consequently, products will be produced which could not have been produced before.

BASIC FUNCTIONS OF PRODUCTION

Before one can clearly assess whether or not particular production operations or processes can or should be automated, the basic functions of the production process — and the stages of manufacturing which incorporate them — should be understood. The four functions associated with any production process are: (1) materials processing, (2) materials handling, (3) control, and (4) information processing (see Figure 2).

Materials Processing. Materials processing operations are those that alter the form or nature of raw materials, components, or assemblies. Examples of such operations include machining, casting, molding, mixing, vulcanizing, and stampin . Included in this category are adjunct processes such as assembly, inspecti n, testing, and packaging.

Materials Handling. This function involves the movement of materials, components, or assemblies within or between the operations of a process. Examples include moving bulk materials via belt conveyors or air systems, conveying parts or assemblies by means of powered or magnetic conveyors, and indexing and/or positioning parts in an assembly operation.

Control. Control operations constitute the initiation, regulation, and termination functions of a manufacturing process. Such control systems run the gamut of sophistication from limit switches and relays to numerical and

Production Stage	Possible Functions Involved			
	Processing	Handling	Control	Information Processing
Raw materials handling & processing	●	●	●	●
Parts production	●	●	●	●
Inspection		●	●	●
Assembly	●	●	●	●
Test		●	●	●
Packaging		●	●	●
Storage		●	●	●
Shipping		●	●	●

FIGURE 2.   How steps in manufacturing process incorporate basic functions.

computer controls. Here, the key element of automation enters the picture — precise sensing and measuring of process variables — to provide the accurate control of the process that is becoming indispensable today.

Information Processing.   Some aspects of information processing are a part of the control function, as in the case of process control computers. Indeed, all information processing is technically a control function, in that it is ultimately aimed at better business or process control. However, the direct control involved in a system of relay circuits or computers is much different than an information system designed to indirectly improve control via increased knowledge about the process. An information system need not be "connected" to a process at all, in the sense that what is learned is immediately fed back to the process in some regulatory fashion. In more sophisticated applications, however, information systems may feed through computer links to directly regulate a process.

In this text, the distinction between direct and indirect control will be preserved by considering control and information processing as separate functions. The control function will refer to the direct hardware regulation of processes, while information processing will refer to the accumulation and evaluation of process data so as to better understand and thereby improve process productivity.

STEPS IN THE MANUFACTURING PROCESS

Every manufacturing process — hence every opportunity for automation — incorporates these fundamental functions. While the particulars of how these functions are implemented in a specific process vary according to the special requirements and circumstances involved, there are certain general production steps that are common to all manufacturing processes. These steps are

1. *Raw materials processing.* Whatever the end product may be that is being manufactured, specific raw materials are necessary. The first stage in any manufacturing process involves the handling and processing of raw materials.

2. *Parts production.* After raw materials are processed, components and/or products must be produced.

3. *Inspection.* To assure that the components and/or products are of the quality necessary, inspection and other quality control operations must be performed.

4. *Assembly.* Once all component parts for a product pass inspection, they must be combined, mixed, or assembled into a final form.

5. *Test.* At this point the completed product must be tested for performance, inspected for final quality, and measured against any other product requirements.

6. *Packaging.* Once finished products are tested and inspected, generally they must be packaged. This operation may involve merely placing the product in cartons, or the procedure could be much more complicated where fragile or hazardous products are involved.

7. *Storage.* Ready-to-ship packages must be moved to a warehouse, stored in an easy-to-find manner, and retrieved quickly to fill orders.

8. *Shipping.* Access from storage to the shipping area must be efficient and orderly so that shipping vehicles will not be kept waiting and orders will not be lost or mixed up.

STRATEGIES OF AUTOMATION

Analysis of these eight stages in the typical manufacturing process shows that three of them — raw materials processing, parts production, and assembly — involve the materials processing function; whereas all of the stages could involve the handling, control, and information processing functions. It is this overlapping of basic functions throughout every stage of production that makes automation such a natural course in the improvement of manufacturing operations and processes.

Any one or all of the basic steps in a manufacturing process can be automated. Of course, the degree to which any operation or process should be automated will depend upon what will be allowed by the existing manufacturing technology, labor availability, materials, product design, and business economics. Fundamentally, there are two interrelated strategies for automation. The application of one inevitably leads to a consideration of the other. The two strategies are: *single-operation automation* and *integration of process operations*. A simple example will illustrate the relationship between them. Imagine that a manufacturer determines that he is losing money by manually packaging his product. Not only is he paying workers more than the operation is worth, but the slow rate of packaging constitutes a bottleneck which unduly restricts the amount of finished product that can be shipped to customers. The market demand is far greater than his capacity to produce, and the packaging operation is identified as the principal cause.

The manufacturer decides to automate a materials processing operation. The many manual operators are replaced by an automatic machine system and one supervisor. After installation and debugging, the machine turns out the product at a high rate. Is the problem solved? Probably not, because the operations preceding the machine were geared to manual operations. It may not be possible to feed the new equipment at the rate necessary to assure its profitable operation. How about subsequent handling operations? They were not designed for the increased output of the new operation either. Also, will the means to handle the product between operations be able to keep pace under the new conditions?

Semiautomatic or automatic machines cannot perform without a continuous supply of materials or parts to be processed, and a continuous removal of the processed products. Thus, when an automatic machine is introduced into a production process, numerous — perhaps all — other operations must be adjusted to the new rate of the mechanized procedure.

The balancing and improving of operations fore and aft of an automated operation forces a manufacturer to think of the process *as a whole*. He begins to perceive the *interrelatedness* of the various operations involved, and soon he will be designing integrated systems that will yield orders-of-magnitude better results than if he were to continue treating each operation as an isolated component. In any enterprise, economic necessity compels the progressive automation of production operations which, in turn, fosters the growth of integrated systems. The degree and rate at which this evolution occurs will depend upon the market for, and competitive standing of, the particular product involved.

The principal integrating function in a system of automated production is control. Production equipment, handling systems, and information monitoring stations — all must be controlled. Thus, the integration and centralization of automatic control systems is a must for any automated process. In this way, *sequences* of operations can be initiated, regulated, and terminated according to an overall systems plan.

1. Identify and define the existing problem.

2. Outline desired objectives.

3. Collect information concerning all areas, i.e., production requirements, costs, product design, pertinent to the automation project.

4. Generate various approaches that would achieve the results sought.

5. Define the system capabilities that would be necessary to implement each approach.

6. Develop standards with which to assess the design concepts, including such things as how the new system or systems will interface with existing operations, total investment required, return on investment, applicability to long-range production and enterprise goals, etc.

7. Decide on one approach.

8. Specify the equipment and components that will best implement the chosen concept.

FIGURE 3.   Attacking a production problem with automation.

IMPROVING EXISTING OPERATIONS

Deciding where, when, and how extensively to automate a process is not easy, (see Figure 3). Basically, there are two types of circumstances in which the automation of manufacturing operations should be considered. The first circumstance is when it is necessary and/or desirable to upgrade existing operations and/or processes. The second circumstance is when a new process is being planned as a whole, as when an entire process is being automated or when a plant is being built from "scratch" to manufacture a new product. The latter situation presents a better opportunity to do a more systematic, comprehensive, and profitable job of automating. In fact, many present-day products would be unmarketable if their manufacture was not planned with automation in mind, and executed in highly automated plants.

However, one need not tear down his old plant and build a new one to reap the many benefits of automated production. Close scrutiny of existing operations will reveal numerous opportunities to apply the principles of automation technology to increase product volume, cut costs, improve quality and, in general, to increase productivity. The following paragraphs focus on some of the things to do and look for when seeking to upgrade existing operations via automation technology.

1. *Determine process flow pattern.* The first thing to do is to look beyond the details of individual operations and identify the basic flow pattern of the overall process. Process flow patterns can be broken down into three basic types: (a) one material in, one product out; (b) several materials in, one product out; and (c) one material in, several products out.

In the first case, essentially the same material that enters a process goes out as part of the finished product. The "straight-through" processing of many bulk materials are examples of this type of flow pattern. In the second case, a lot of different materials and/or subproducts enter a process; each of them undergoes its own particular set of processing operations; and the various elements are ultimately assembled into one final product. The manufacture of appliances, automobiles, etc., provides pertinent examples of this type of flow pattern. In the third case, one material enters the process; the material is broken down into several components which undergo separate processing; and several distinct finished products emerge. The production of gasoline, motor oil, and kerosene from crude petroleum is an example of this type of process flow.

The straight-through — "one material in, one product out" — type of process flow is easier to "fully" or partially automate than the other two, assuming the technology exists to do the job. The logistics of successive, one-after-another, operations is much simpler to analyze, implement, and maintain than a complex of several interrelated lines that branch off and/or funnel into one another. There are fewer possible problems to guard against with the straight-through process, and fewer process alternatives to provide for. Also, it is quite a bit easier to balance the production of a straight-through line after an operation or two has been mechanized.

2. *Look for operations in which the content of repetitive labor is high.* Whenever a worker or workers are doing the same identical task over-and-over again, the operation involved is a natural for automation. Machines can do repetitive tasks faster and better than humans — and machines won't get tired or bored. It doesn't matter whether the operation presently requires skilled or unskilled labor, it can probably be automated. The result will not only be more and better products at lower cost, but a better, more profitable utilization of manpower as well.

3. *Look for operations with high material throughput.* It is much easier to automate an operation in which the equipment setup stays the same and the material or products processed are similar. Every time automatic equipment is interrupted, time and money is lost. In general, the greater the throughput, the better opportunity there is for automation.

4. *Look for operations which involve sequences with similar time cycles.* Any time several successive operations are accomplished in about the same amount of time, there is an excellent possibility that they can be combined sequentially in a single automated machine or system. Many existing automatic assembly

machines, for example, are mechanized integrations of what were once several discrete operations accomplished by many workers and/or a variety of tools and/or equipment.

5. *Look for operations which have similar levels of automation.* It may be that such operations are still interrupted by manual procedures. If so, eliminate or mechanize them so that total process flow can be increased toward the rate at which its mechanized operations function. Also, many times operations with similar levels of mechanization can be grouped together and integrated, gaining another order-of-magnitude improvement.

6. *Look for ways to improve the design of the product.* Oftentimes, slight but significant modifications in the product will make its manufacture much more amenable to automation. Products that were designed with conventional techniques in mind can sometimes be altered so as to make them more easily handled and processed automatically.

7. *Look for ways to reorganize the existing sequence of operations.* Many production processes grow up piecemeal, under the pressure of day-to-day expediencies. Evaluating a process as a whole may suggest new ways of organizing production operations so as to evolve new production sequences that would be more conducive to mechanization and/or integration.

PLANNING A PROCESS AS A WHOLE

Attacking the automation of a process from scratch requires analyzing every element involved. To be efficient and successful, automated processes must be more than an assemblage of interrelated equipment. They must be skillfully managed. Planning must be careful, complete, and competent. Implementation of the plans must be expert. The following paragraphs will briefly outline the considerations which must be fully examined and evaluated when planning a comprehensive automation program.

The Assembly Viewpoint. A product that requires assembly by automation must have component parts of high quality, high uniformity, and reasonably fine tolerances. Parts of nonuniform character and widely varying material properties must be assembled by hand with individual fitup where and as needed. Thus, the production methods used are critical. In order to solve the assembly problem profitably some parts must be produced in the machines as and when needed to insure maximum control along with high operational reliability.

Invariably the right process must be selected to result in lowest production cost overall considering all factors that can influence final cost. While it is possible to produce antifriction bearings by turning and grinding rings cut from tubing which then move to automatic gaging, selection, and assembly, it is far more efficient and profitable to produce the rings by cold forming rather than turning. The process is far faster and thus more efficient since fewer machines are needed to match the necessary production rates of an automated system. In addition, the raw stock is far less costly. One might conjecture here that the next economic step could well start with continuous compressing and sintering of the rings from suitable mechanical alloys of powder.

Process Combinations.   Basic to the concept of automation is the combination of a series of production operations into a continuous or automatic batch-type system. As a corollary it is obvious that the most profitable development of an automated system should look toward minimizing the number of steps required. The fewer needed, the lower the cost of the equipment and the shorter the process cycle.

To actually begin any automation planning properly, the fundamental approach used as a framework should be conceptually that of assembly. A large proportion of the products produced are or end up as assemblies of components. If the processes cannot be arranged so as to produce the final assembly, it may be possible to produce subassembly modules.

Wherever feasible, the maximum number of operations possible should be done before the product is released. Real economies are to be gained when process steps can be combined into one or into a single functional unit. Thus, inserts can be included in plastic molded or die cast parts to permit direct transfer to final assembly by automatic arm.

Processes Demanding Automation.   Some processes and production methods are suitable for purposes that are not demanding, but when rigid qualifications arise they are beyond controlling manually. To attain exacting tolerances in ultrathin gages of steel — say plus or minus 0.0001-inch on 5 mil strip — automation is a must. Sensitive control to such limits today can best be obtained profitably by means of computer controlled rolling mills.

Modern chemical processes as well as refining operations, gas transmission, and a growing variety of metal cutting operations fall into the same category. Precise automatic control may render the process not only more precise and useful but, from a profit standpoint, also produces more saleable output from the raw materials input. The loss in steel from a mill that rolls to plus or minus 0.002-inch vs. one that holds plus or minus 0.0001-inch can reach as much as plus or minus 8 percent when 24-gage stock is running. This means that between the high and low limits the possibility of a 16 percent total variation can exist: footage from a 75-ton roll of 24-gage steel can vary 24,000 square feet; conversely, by the measure of exact length the weight can vary a total of 12 tons. Obviously, fast, continuous processes must be accurately controlled via automation techniques if the customer is to be given a quality product with guaranteed conditions and the manufacturer is to insure a fair profit return on the operation.

Plant. Will a new facility be constructed or an old one updated? While new facilities are not necessary, it is important that the building to be used be in good condition so it will last for many years. A substantial amount of money should not be spent renovating a building that would have to be replaced in a few years. Furthermore, the plant must be suitable for the job to be done. It should allow for the kind of flow pattern envisioned and permit the volume of production that is anticipated for many years to come.

The age and condition of the plant is also important from the standpoint of whether or not its facilities are sufficient for the greater power, sewage disposal, water, and other requirements of automated production. Also, the architecture of the building is extremely relevant to the kind of process involved. Single and multistory structures lend themselves to different kinds of production. An automated process is less likely to be successful if it is forceably fit into an unsuitable plant design. In general, the single-story building lends itself better to automation because of the accessibility and ease of delivery of raw materials into the process, plus the ease of transfer from the end of the line to storage and shipping facilities.

REFERENCES

Automation in Business and Industry by Eugene M. Grabbe, John Wiley & Sons, Inc., New York, 1957.

Automation Systems, Conference of the Electronic Industries Association held at Arizona State College, Engineering Publishers, New York, 1958.

Technological Change, A Series of Eight Reports by Corplan Associates of Illinois Institute of Technology Research Institute, 1964.

Industry Case Examples

CASE 2A

Telephone Cable Production

Automated production of wire and cable is well illustrated by the work of the telephone industry. The year 1970 saw new records set for production of telephone cable and wire, Figures 4 and 5. More than 250 billion conductor feet of exchange cable were made. This, in addition to some 53 billion conductor feet of purchased cable, would form a single wire 58 million miles long.

Because of this demand and the limits set by present production methods, the Engineering Research Center of Western Electric developed a new process for the continuous production of aluminum wire by extrusion, Figure 6.

A prototype production machine is being tested at present. Designed to extrude aluminum wire direct at speeds approaching 60 mph, it is expected to begin a new stage in the automation of wire production which now is drastically limited by conventional wire drawing techniques originally developed by the Egyptians four thousand years ago.

Telephone drop wire is produced by continuously electroforming a copper layer over a steel wire core at speeds of 100 feet per minute on 25 wires simultaneously and continuously under automatic control.

CREDITS: Western Electric Manufacturing and Supply Unit of the Bell System.

FIGURE 4.   Wire reel supply end of continuous cable-making machine.

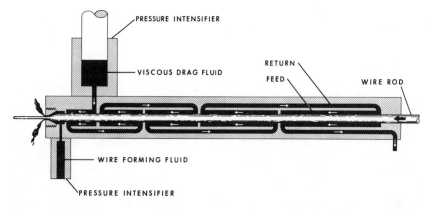

FIGURE 5. Cable-twisting head of cable-making machine.

FIGURE 6.   Wire rod is fed into the continuous hydrostatic extruder and dragged through the high-pressure chamber by a viscous fluid such as warm beeswax. The multiple chambers create the proper stress and pressure relationship. Actual deformation of the rod is done by forming fluid in the die, so that there is no metal-to-metal contact. Intensifiers (left) inject the viscous-drag fluid and the forming fluid at high pressures. (Photograph from Western Electric, Manufacturing and Supply Unit of the Bell System.)

CASE 2B

Automatic Welding

Welding has always been a major operation in automotive body construction. In designing the Vega automobile, General Motors engineers took great pains to consider all of the methods that might be employed to simplify welding. In addition to automatic welding tools on the market, one of the commercial tools employed is Unimate, shown in Figure 7 with welding guns. But these amazing manipulators cannot reach all locations, so the body was designed to accommodate these tools. The joints were made as accessible as possible, and in plane, when possible, so that a minimum number of manual welds could be applied to first hold the body together. Then, the Unimate and other automatic welding devices could be used to complete the welds. Approximately 95 percent of the welds in the body, and its components, are performed automatically.

The quarter-panel inner and wheel-house assembly is automatically built-up at another location. From here, these parts are conveyed to what we call the side-framing area. Here basic side components of the body — the hinge pillar and quarter outer, rocker, and quarter inner panels are loaded into a building fixture which accurately locates and holds the components. They are shuttled through a series of automatic spot welders where the precise location of welds is again mechanically assured.

The basic major subassemblies are now ready to start their journey down the continually moving assembly line.

First, the underbody is automatically loaded from its bank onto the moving building truck, which is actually a very precise positioning device. Then the side frames are automatically moved into position and accurately locked in place on the sides of the building truck. The complete roof structure is sequenced from an overhead storage bank, automatically dropped in place, and clamped in the proper location.

Enough welds are placed in strategic areas to secure the body structure for dimensional accuracy. Some of these welds are made manually with portable guns, but most are made with automatic welders. A two-stage automatic tack welder in this area indexes to the moving body, makes 65 welds per side, and retreats. This machine accomplishes its welds in 4.2 seconds, and completes its total indexing, welding, and withdrawal cycle in 11.5 seconds.

Additional automatic welders complete the body framing operations and the side frame fixtures are automatically removed. The body on its building truck leaves the continuously moving line and moves onto a special conveyor system designed to accommodate a battery of Unimates.

Unimate is just what its true name implies — an industrial manipulator — a machine that will move a tool through a predetermined series of motions with amazing accuracy and repeatability.

At Lordstown, the Unimates are equipped with specially-designed lightweight portable welding guns. An operator uses a "teach cable" to initially move the welding gun through its desired cycle. Upon command from the operator, the Unimate memorizes the exact "point-in-space" location of its tool, and the series of motions required to move it there. It can store up to 180 commands in its memory system. When called upon, the Unimate can accurately repeat that pattern of tool movements — time after time.

Along each side of the line, eleven Unimates and one automatic welder complete body welds with a consistent accuracy of 1/16 of an inch. Though they are capable of making up to 60 spot welds, time limitations restrict each Unimate to a maximum of about 20. Including the side frame operations, Fisher Body Lordstown assembly now operates a total of twenty-six Unimates, with two spare units as emergency replacements, as well as forty-nine other automatic welding devices.

The manipulators are not capable of operating along a moving assembly line. Therefore, we developed an unusual eight-station shuttle conveyor and locking system that automatically indexes the body-building trucks and holds each one stationary at a precise location in every station.

One 20 horsepower mechanical drive moves the eight bodies and their building trucks. A total load of 40,000 pounds indexes accurately in 8 seconds — accelerating to 36 inches per second in 15 inches — and decelerating at the same rate.

As another part of the quality control program, self-contained memory systems monitor the function of each on-line Unimate. Should a machine fail to complete its assigned program, a signal light appears on a master control board above the line. An audible signal alerts the maintenance force to investigate and correct the problem.

CREDITS: Chevrolet Motor Division, General Motors Corporation.

FIGURE 7. Automatic welding equipment places 95 percent of the 3,900 welds on each Vega body.

CASE 2C

Computerized Soup Production

Up to 15,000 gallons of soup an hour can be made by a new $840,000 automatic materials handling system installed at the Heinz Kitt Green factory in Lancashire, England.

The computer-controlled system, first of its kind in this type of food manufacture, selects the raw materials for forty different soups in the required quantities and combinations from stocks of forty-four separate ingredients — eighteen held in powder form (flours, sugar, etc.), eighteen pulse type ingredients (peas, beans, barley, etc.), and eight liquids. It then controls their automatic delivery to the cooking point.

Five different production lines are served by the system, three for soup manufacture and two for a range of spaghetti varieties. A batch is delivered to one of twelve mixing vessels every two minutes.

Providing an efficient, hygienic, and economic solution to the problem of handling ingredients in the large quantities demanded by new manufacturing processes, the system ensures a high degree of product consistency. In every batch of 500 gallons each ingredient is measured to an accuracy of ± 2 percent; most are within ± 1 percent.

The system is the result of a joint study by the Heinz engineering and research divisions and subsequent collaboration between Heinz and Henry Simon Ltd., the main contractors. Great care was taken to ensure that standards of quality and hygiene were fully maintained. Control panels and manually operated sections were designed so that plant operators could work in the best possible conditions with minimum effort.

Twenty men headed by two specially trained computer technicians operate the plant. All the manufacturing facilities serviced by the system were new and did not therefore have an existing labor force; but approximately seventy additional men would have been required to make the same quantity of soup by traditional methods.

The whole plant is controlled by a CON/PAC 4040 computer system supplied by GEC-Elliott Process Automation Ltd.

Sequence of Operations: Dry materials are tipped into seven sack tipping units as called for by the computer controlled panel on each unit. The materials are seived and each is directed to the appropriate storage bin in a nest of 36 bins shown in Figure 8. They hold an average of four tons each.

Each bin is equipped with one or more volumetric dispensers. On demand from any mixing vessel, the computer starts up the dispenser drives which run until the computer calculates that the required quantity has been discharged.

This is done on a time/volume basis.

The dispensers discharge into one of three pipes through which the material is blown using 1200 cfm of free air at 35 psi. The ingredients are routed to the correct vessel by a system of multiport motorized valves and are separated from the conveying air by equipment housed in a penthouse over the mixing area.

From the penthouse, the batch is then dropped to the mixing vessel at the cooking point.

Additionally, ten liquid-storage vessels are connected by a system of pumps, filters, flow meters, valves, and stainless steel pipework to the mixing vessels. At the same time as the dry materials are being delivered to these vessels, the computer activates the liquid system and ensures that accurately measured quantities of liquids are fed into the appropriate vessel at the correct time, as shown in Figure 9.

When all the ingredients are in the mixing vessel, the computer starts the mixing process and runs it for a predetermined time. On completion, a light indicates to an operator in Figure 10 that the soup is ready for a final quality examination before being transferred to the next part of the manufacturing process.

The Computer: To carry out the control function, the computer in Figure 11 uses a core store of some 16,000 words of 24 bits each. In this "memory" is held all the information needed to produce any of the range of forty soups on any of the three soup lines, and any of three sauces for the spaghetti lines.

Any malfunction in the plant is immediately and automatically detected. Every twentieth of a second the computer scans a signalling system of 900 inputs. If an abnormality is recorded, a visual signal is given at the appropriate location and a typewritten statement of the nature and location of the fault is made in the control room. Depending on the significance of the fault, the plant is automatically shut down in whole or in part until the error is rectified.

In addition to the control function, the computer also assembles operational data, such as stock levels in bins or the number of batches produced in a shift. This information is available automatically at the end of a production run or at any time on demand.

Cleaning: A vital part of soup processing is the proper cleaning of equipment at regular intervals. Provision is made in the computer program to initiate automatically a purging of the dry system and a full in-place cleaning of the liquid system at the end of every shift.

Plant Performance.   With overall capacity of 30 x 500 gallon batches per hour producing up to 15,000 gallons of soup per hour, the plant is comprised of three main sections:

1. *Powder system* (flours, sugar, salt, starches, semolina, milk powder, spices). Eighteen raw ingredient materials used in quantities ranging from four ounces to 600 pounds per batch. Up to fourteen ingredients in any batch. Maximum batch size 1,000 pounds. Minimum dispensing accuracy of ± 2 percent on all materials. Most materials are dispensed to about ± 1 percent accuracy. Throughput 30 batches per hour. Average material usage six tons per hour.

2. *Pulse system* (barley, lentils, dried peas and beans, short pastas). Eighteen raw ingredient materials used in quantities ranging from 10 pounds to 800 pounds per batch. Up to fourteen ingredients in any batch. Maximum batch size 1,000 pounds. Minimum dispensing accuracy of ± 2 percent on all materials. Throughput 30 batches per hour. Average material usage three tons per hour.

3. *Liquid system* (water, milk, cream, tomato paste, melted butter, vegetable oil, chicken stock, protein solutions). Eight raw ingredient materials used in quantities ranging from four pints to 450 gallons per batch. Up to seven ingredients in any batch. Maximum batch size 450 gallons. Minimum dispensing accuracy ± 2 percent on all materials. Most materials are dispensed to about ± 1 percent accuracy. Throughput 30 batches per hour. Average material usage 12,000 gallons per hour.

CREDITS: H. J. Heinz Company Limited.

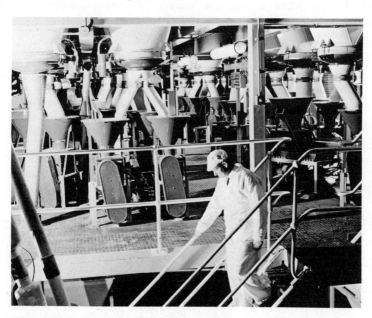

FIGURE 8. Materials handling system at Kitt Green. The bin nest area, showing the dispensers on the middle floor. (Photograph from H. J. Heinz Company, Ltd., Hayes Park, England, 1970.)

FIGURE 9. Complete array of mixing vessels at Kitt Green shows how dry ingredients are computer-fed through large pipes from ceiling, while liquids enter through smaller pipes around the mixers. (Photograph from H. J. Heinz Company Ltd., Hayes Park, England, 1970.)

FIGURE 10. Materials handling system at Kitt Green. The mixing tank area showing the main indicator and control panel for the mixing operation. (Photograph from H. J. Heinz Company, Ltd., Hayes Park, England, 1970.)

FIGURE 11. Heart of the automatic soup production line is this computer room. Instrument with TV-like screen keeps constant check by computer. Typewriter is used for print-out of messages from production line. (Photograph from H. J. Heinz Company, Ltd., Hayes Park, England, 1970.)

3

Handling
Materials in
Process

As outlined in the previous two chapters, every manufacturing process involves handling. Even a process consisting of a single operation must have raw materials brought in and finished products taken away. The more complicated the process, the greater are the number of handling procedures that will be required and, therefore, the greater are the potential benefits of automating those procedures.

Since handling procedures connect two or more production operations, all efforts to automate handling must be geared to the desired production rates of the operations involved. The overall goal when automating handling procedures is to provide the mechanized transfer links between process steps that will result in the optimum performance of the process as a whole.

BASIC TYPES OF HANDLING PROCEDURES

The problems encountered when automating the handling function — and the character of the resulting systems — depend upon two general considerations:

first, the type of handling procedure required; second, the nature of the material involved. There are five basic types of handling procedures:

1. *Movement of materials to a process.* Every production process begins with the transporting of specific materials and/or components to initial process operations. These materials and/or components may be iron ore, coal, gravel, etc., from mining facilities or rough or finished parts from other manufacturing plants that require further processing and/or assembly. Systems used to perform this type of handling procedure vary from belt and pneumatic conveyors to pressurized piping systems.

2. *Movement of materials between operations.* The tying together of production operations with automated handling systems depends upon the kind of material or the shape of the part being moved, the rates at which materials or parts must be supplied, and the kinds of obstacles (if any) that must be overcome. Belt, screw, chain drag, roller, and magnetic conveyors are used in this type of handling procedure as well as air chutes, tote carts, industrial robots, etc.

3. *Manipulation of materials within operations.* Parts on their way to a turning operation must be oriented properly and placed in position accurately before cutting can begin. The tiny components of a complex assembly must be oriented, fed, and placed at the proper rate. Packages must be accumulated and oriented before they can be automatically palletized. These typical examples illustrate the diversity and complexity that can be involved in automating handling procedures within particular production operations.

4. *Movement of materials away from a process.* After processing and packaging operations are completed, finished products must be moved to shipping or storage areas. End-of-line handling procedures and systems may be very similar to or very different from the raw materials handling systems used, depending upon the degree to which the form of the incoming materials change via subsequent processing operations. The enrichment of iron ore would still result in the movement of ore-like material at the end of the enrichment process. On the other hand, packaged ball bearings are nothing like the steel stock from which they were produced. Hence, the end-of-line handling in a ball bearing manufacturing plant would be nothing like the systems used to input raw materials. Various types of conveyors, tow carts, etc., are used for end-of-line handling.

5. *Movement of materials into, within, and out of storage.* Automation of warehousing has taken great strides in the last ten to fifteen years. Warehousing used to be the last part of a process to be improved. If a manufacturer fed his product to storage via conveyor, he thought he had an advanced storage system. Today, of course, this perspective has changed. As the rest of the production process was automated, warehousing emerged more and more as a bottleneck.

Once it became possible to turn out a product in mass quantities, it became crucial to maintain a proper level of inventory, know the location of each type of product, and move the product in and out of storage quickly. Advent of mechanized and computer controlled warehousing systems has brought the available level of warehousing automation up to that possible with other stages in the manufacturing process. Factors influencing the degree of difficulty of warehouse automation include: quantity and variety of products, possible hazards encountered in handling the products, delicacy or break-ability of products, and the environmental conditions necessary to maintain the integrity of products.

The processing steps necessary to produce a product dictate the types of handling procedures required. The more complex the product is, the more production operations there will be and, consequently, the greater the number and variation of handling procedures that will be needed. The straight-through production of textiles, for example, involves much simpler handling procedures than the parallel processing lines involved in manufacturing portable radios.

How each type of handling procedure is accomplished within a particular production context depends, in general, upon the form of the material being handled. Basically, there are three forms of materials pertinent to manufacturing: bulk materials, discrete parts, sheets and webs. The following sections focus on how the characteristics of these forms of materials influence the type and difficulty of the automation employed.

BULK MATERIALS

Bulk materials include dry bulk ores and products, liquids, and slurries. As raw materials or finished product, they can be handled in a continuous flow (as with conveyor belts, pneumatic systems, etc.) or in batch form (as with cars, carts, trucks, etc.). In-process handling of bulk materials involves such procedures as transporting, routing, feeding, and dispensing powders, liquids, and other dry and liquid bulk products.

Regardless of the application, successful automation of bulk materials handling depends upon determining the characteristics of the material, selecting the proper handling equipment, and applying the equipment to the particular process requirements (see Figure 1).

Material Characteristics. Where material flow is integrated with processing, the success of an entire process may depend upon the performance of one or more bulk-handling systems. Changes in the behavior of the material before, during, or after transfer can affect system performance. Thus, materials are the key elements in bulk materials handling systems.

Conveyor Type	Suitable Material
Belt	Can handle most materials, at high capacities, over long distances, up and down slopes.
Apron	Best for heavy, abrasive, or lumpy materials.
Flight	Pushing action of this equipment is conducive to handling nonabrasive materials.
Chain Drag	Good for light materials, coarse or fine.
Screw	Fine or moderately sized materials.
Oscillating	Can handle hot, stringy, abrasive, or irregularly shaped lumpy materials.
Skip Hoists	Any bulk material.
Bucket Elevators	Some types are suitable for free-flowing, fine to medium-size lumpy materials; others are best for difficult-to-handle materials (such as materials with large lumps, fluffy materials, etc.).
Vertical Screw Elevators	Will elevate many of the materials handled successfully by screw conveyors.
Pneumatic	A variety of bulk materials capable of suspension in an air stream.

FIGURE 1.   Matching conveyors to bulk materials.

Questions to be answered about the material to be handled include: Is it dry, liquid, or sticky? What is its chemical composition? What percentage of moisture can be tolerated? What is the range of particle size of the material? What is its density? Are there any special characteristics — such as whether the material aerates, is hot, corrosive, degradable, explosive, toxic, or temperature sensitive — that should be noted?

Where dry solids are involved, the following material characteristics are most important:

1. *Particle size.* Constituent particles of a bulk materal can range in size from very fine to granular, lumpy, and irregular.

2. *Abrasiveness.* This characteristic is related to the shape and hardness of the particles. Usually, materials that are hard and/or angular in shape are the most abrasive.

3. *Flowability.* The flowability of a material is influenced by such factors as the size and shape of the particles, and the moisture content of the material. A good indication of flowability can be obtained by measuring the "angle of repose", i.e., the angle formed between the surface of a normal, freely fed pile of material and the horizontal plane.

4. *Compressibility.* All materials are, to some extent, compressible. Vibration and/or pressure will compress various materials anywhere from 15 percent (as with table salt) to 200 percent or more (as with nitro-cellulose).

5. *Uniformity.* Materials may segregate in storage bins so that by the time they are fed they are in the form of particles of a uniform size or particles of well-dispersed mixtures.

6. *Cohesiveness.* This characteristic refers to the tendency of like molecules of a material to hold together. The greater the cohesiveness, the greater will be the measured angle of repose.

Some of the characteristics that should be kept in mind when automating the handling of slurries or liquids are:

1. *Composition.* Many times slurries cause flow control problems because the solids tend to settle out in valves or fittings. The tendency of a slurry to "settle out" in this fashion must be determined before trying to utilize metering or gear pumps, for example.

2. *Viscosity.* This is the property of a fluid or substance that resists internal flow. Flow control problems can result with liquids or substances that have very high or changeable viscosities.

3. *Foamability.* Whether or not a liquid, slurry, or substance will foam under pressure or movement is crucial to the kind of system that must be employed. Highly foamable liquids, slurries, or substances result in widely varying densities which could play havoc with the functioning of the handling system.

Selecting the Proper Equipment.   The fundamental question to be answered here is: What kind of equipment is best suited to handle a specific material in the manner required? Where the movement of large quantities of bulk materials is required, for example, conveyors are the most appropriate systems. But what type of conveyor?

Pneumatic conveying systems utilize blown or drawn air to force or suck bulk materials through directional pipes. However, pneumatic conveying is only suitable when the material to be conveyed is composed of elements that are less than three-inch cubes in size, free flowing, reasonably dry, and not easily abraded. Also, process flow capacities should not be more than 400,000 pounds per hour, and the conveying distance should be less than 4,000 feet.

Where mechanical conveyors are concerned it is necessary, first, to determine which of the "carry", "push," or "drag" conveying actions of various equipments is necessary for a specific material. For example, whereas most materials can be "carried" effectively by a belt conveyor, an unusually free-flowing material may require the "pushing" action of a bar-flight or screw conveyor. Certain materials may require the attachment of special carriers (such as pans) to a standard conveyor for effective movement. Fluffy, sticky, hot, highly abrasive, and environmentally sensitive materials present their own unique problems in regard to conveyor selection and use.

Other criterions that should be applied to the selection of system equipment include:

a. The system should be safe. Not only must there be no hazards for workers, but the integrity of the material being handled must be maintained.

b. It should be easily adaptable to the layout of the plant to be used.

c. Initial and installation costs should be as low as possible without compromising production requirements.

d. The system should be easy to maintain.

e. It should be amenable to automatic control — either because the system will be installed with automatic controls or because it may be necessary and/or desirable to make it automatic sometime in the future.

Applying Equipment to Process Requirements.   Determining the characteristics of the material to be handled and selecting the equipment to be used are, together, only half the battle of automating handling operations. The most important step remains, i.e., integrating handling operations into process operations so that the desired production requirements can be met. There are four basic questions to be answered when designing an integrated bulk materials handling system:

1. Where must specific amounts of material be located at specific times to satisfy individual processing operations?

2. What amounts of material, by weight or volume, must be transported at certain times to be at the designated locations when needed?

3. In what directions must material be transported to arrive at those locations? (Must the material be elevated at any point? Will it pass through hostile

environments? Will there be any hazard to workers? What is the most effective layout?)

4. At what speeds must the material be moved to specific locations so that processing rates will be satisfied? (There is no more certain way to foul up a potentially profitable automated process than by mismatching production and handling speeds.)

DISCRETE PARTS

Discrete parts include such varied items as microcircuits, fasteners, and castings. Large or medium sized parts may be moved by flat conveyors, overhead conveyors, etc., and may be transported uphill and/or around obstacles via chain or magnetic conveyors. Small or miniature parts can be automatically handled, oriented, and fed by a great variety of conveying and/or feeding systems.

In all the possible applications involving the automation of discrete parts, however, potential problems to be studied and solved fall into two general categories: first, those problems which are intrinsic to the part or piece, i.e., the design or characteristics of the part that may affect handling; second, those problems associated with system design, i.e., the influence of the part upon satisfying the various feed rate, orienting, and positioning requirements of the process.

Problems Intrinsic to the Part. Design and condition of the part are very important to automatic handling. At some time before design commences, the basic *part* design should be frozen. It will do no good to design and build an elaborate handling system around a particular part configuration, only to change the part in such a way that the system developed is no longer appropriate. On the other hand, *slight* modifications of the part might make the design of an automatic handling system much simpler. Perhaps the shape of the part can be altered. Or, maybe a less environmentally sensitive material can be substituted for a temperamental one without compromising the end-performance of the product.

Problems associated with the conditions of the part include:

1. Variable tolerances can cause a high part-rejection rate.
2. Unwanted burrs on metal parts — or the presence of oil, chips, or other foreign materials on or between parts — can cause jamming of parts and/or clogging of conveyors, feeders, or escapements.
3. Severe handling procedures — such as tumbling, clamping, or rolling — can impair the integrity and usefulness of the part.

Oddly-shaped, bulky, or miniature parts or pieces present special problems in the design of automated handling systems. In the cases of oddly shaped and/or bulky parts, for example, specialized equipment is needed to perform the transport, positioning, transfer, and/or placing procedures. Often, one or more of these procedures is performed manually until production volume and/or costs can justify the expenditure required to create the unique carriers and manipulators required.

The special problems associated with handling miniature parts center around the following considerations:

a. The smaller the part, the greater the chance that irregularities in shape will be disproportionately large relative to the basic part-size, thereby creating feeding difficulties.

b. Smallness of such parts creates a "weightless"-type condition which makes the parts over-react to standard power drives.

c. External conditions affecting the parts or the components of the feeder which might be ignored with regular-sized parts often create poor or on-again/off-again feed performance when very small parts are involved. Such conditions include a buildup of static electricity, high humidity, electromagnetic effects, fallout of dust or other materials usually suspended in the air.

Problems Associated with System Design.   Once the part design has been set and the part condition determined, system parameters must be determined so that equipment can be selected and integrated effectively into the various stages of the production process. The key questions to answer in this regard are:

1. What supply rate is required? Production, scheduling, production rates of related parts, and times required for other simultaneous operations of the process are prime considerations here. Conveyors, feeders, and/or positioning mechanisms must supply parts at rates compatible with the index or cycle rates of the machines involved. Multiple feeding systems should be used where a single system is barely keeping up with the machine it is feeding.

2. Is orientation necessary? In most cases, parts or pieces must be presented to production machinery in specific orientations. The more of this orientation that can be accomplished before the part or piece gets to the machine, the less will special handling be required just before or within the machine itself. In many applications, the position of a part or piece that is best for feeding is not the same as the position that is best for placement. In such cases, orientation is mandatory.

3. Will a checking station be required? In many applications, it is desirable to include a checking device that will signal or stop the production machine when a conveyor, feeder, orienter, or placement device malfunctions. For example, if a transfer device fails to present a part at the right time, continued operation of the production machine can result in costly repairs. Or, feeder malfunction

could result in countless assemblies with missing parts — again, necessitating a costly correction.

4. What type and size of supply system are necessary? Factors influencing the size and type of conveyor and/or feeder include the size and shape of the part, the space available, and the supply or feed-rate of the part.

5. Will placement be required? A part or piece must be correctly placed in the processing machine before a particular production cycle can begin. Transfer or placement devices are usually actuated by the production machine. Such devices constitute the last step in bringing parts or pieces to processing operations. If placement (where needed) is not effective, the rest of the handling system will be to no avail. Placement mechanisms usually utilize vacuum holders, magnets, or spring-loaded fingers to hold parts in the placing fixture. When needed, these mechanisms should be designed as simply as possible.

Selection of a power medium to drive the conveyors, feeders, and/or transfer mechanisms required for a given application is an important part of designing for system success. Available drive mediums are pneumatic, hydraulic, and electric power. Factors to consider when selecting a drive medium include:

1. Speed. Where loads are light and speed of movement moderate to high, pneumatic power may be adequate. Heavy loads and low speeds may dictate use of hydraulic power. This drive medium also provides high motion accuracy.

2. Cycle frequency. Pneumatic drives are excellent where cycle frequency is not excessive and loads are reasonable. But when fast cycles and high surface speeds are necessary, hydraulic or electric power, or a combination of all three may be imperative.

3. Timing accuracy. Where accuracy of timing is mandatory, the use of electric power and control is necessary.

SHEET AND WEB MATERIALS

Many materials handled in automated processing lines come in sheet, film, blanket, strip, belt, or web form. Examples include the continuous rolling or plating of ribbons of steel, the production of papers and fabrics in strip form, and the movement of conveyor belts over long distances at high speeds. Continuous processing of web materials necessarily involves automated handling. To assure a successful handling system, thorough study and evaluation is necessary of, first, the characteristics of the web being processed and, second, the problems encountered in moving a web material through the necessary process operations.

Web Characteristics. Overall physical characteristics of the web are relevant to automated handling. Typically, the unit strength of a web is a function of web density. The thicker and more dense the material, the higher its unit strength. The thinnest or least dense web to be processed on a particular line, then, is the most difficult to handle. Thickness and physical character are also related to the "stretch" or "draw" of the web. In general, the thicker and heavier the web, the less stretch it will exhibit when subjected to a running tension.

Overall web width is related to the linear strength of a web of a specific material and, therefore, the maximum tension it can undergo before tearing. The wider a specific web is, the greater the range of tension that can be allowed before impairing the integrity of the web.

Problems in Moving Webs. In any continuous process, webs must be unwound, moved in process, and wound. These operations are usually repeated many different times throughout a particular process line. The two principle problems involved in performing these handling procedures at high speeds are web tension control and web tracking.

Improper control of web tension can cause a variety of difficulties for producers involved in converting, finishing, or printing web materials. For example, irregular tension can tear webs with low tensile strengths. Excessive tension in slitting operations will increase blade wear and can cause blade deflection, seriously affecting the effectiveness of the operation. Excessive tension can also cause adjacent piles of coated material to adhere together. Off-registration, curling, and similar processing problems can be caused by improper tension control. Various sophisticated control systems are available to automatically control web tension and, thereby, assure the effectiveness of process operations.

Another basic problem unique to handling web materials is keeping the web centered while it is being moved. There are two degrees in which this requirement can be applied. Tracking refers to keeping the material from deviating into the sides or supports of a conveyor or off of a tension roll. For example, if a forty-seven-inch wide sheet of paper in running on fifty-five-inch rolls, the strip could deviate four inches on each side without creating a problem. Alignment is a more stringent requirement. It refers to keeping the longitudinal center of the moving web nearly coincident with the longitudinal center of the roll line.

Whereas tracking is always and everywhere necessary when the continuous processing of sheets or webs is concerned, alignment is a requirement where a precision path is necessary — as in coating, cutting, or coiling operations. Lack of tracking and/or alignment results in nonuniform processing, which means low product quality and/or performance.

There are four fundamental factors of system design to be considered in the centering of webs:

1. Roll alignment. If the rotating elements which move the web are not aligned, the web cannot be expected to track properly. Precise initial roll alignment is important, but subsequent alignments are also necessary to compensate for changes in foundation and/or system component positions. Increasing web tension will also tend to eliminate mistracking due to roll misalignment.

2. Roll shape. Rolls are not perfect cylinders to begin with. But process conditions — such as extreme pressure or heat — can further deform the rolls so as to make tracking difficult.

3. Web shape. Sheet or web material that is perfectly straight and/or of uniform thickness is rarely, if ever, attained. Various types and degrees of deformity produce a variety of detrimental tracking effects. The material, itself, can buckle or twist. Poor web shape or condition may make automated handling impossible. It may be that the quality of material produced will have to be improved before subsequent mechanized handling can be instituted.

4. Material distribution. When a web is being used to transport material (as is the case with a belt conveyor), the distribution of the material across the belt has a great influence on web tracking characteristics. Various distributions produce various pressures along the width of the web. Differing pressures produce a condition in which parts of the web have better traction with the rolls than other parts. Thus, mistracking is inevitable.

Beyond precision alignment of roll, the production of higher quality webs, and the uniform distribution of materials on webs, there are other ways to minimize mistracking. If possible, for example, the web can be moved at a slower speed. The higher the speed, the more are any mistracking tendencies accentuated. Or, web tension can be increased. Care must be taken with this approach, however, since too much web tension can result in damage to the web or increased wear. Also, the web can be run in narrower widths. However, this technique can be costly to production rates. Another way to minimize the effects of mistracking is to use batch-type operations rather than continuous processing. In most cases, the production of sheets instead of strips is a step backward, thus of little interest to those aiming at improving the quantity and quality of product via automation.

At best, these techniques are little more than partial solutions; they help solve the problem of tracking at the expense of achieving full system potential. The real answer is to design a system in which rolls are continuously self-centered. Making use of various mechanical, electrical, or photoelectric means to sense web position, some self-centering systems detect misalignments of various kinds and trigger compensating behaviour (such as changes in speed and/or orientation) on the part of web rolls. In other system solutions, the shape of the rolls and/or the distribution of material on them tends to keep the web centered.

The problems of tension control and web tracking — and the techniques for solving them — apply whether webs are being moved in a straight line, up or downhill, or through the twisting stages of a continuous processing line. As with the other forms of materials discussed, successful integration of web handling procedures depends on designing systems which produce the speeds necessary to match production requirements, maintain the integrity of the product, and result in safe operation.

INSURING ECONOMIC SYSTEM DESIGN

Whether particular handling procedures involve bulk materials, discrete parts, or webs, the following basic guidelines should be applied when designing successful and economical automated systems:

1. Materials handling and storage facilities should be designed for maximum overall efficiency. Do not consider computerizing until a thorough flow study has been made.

2. Integrate as many handling procedures and activities as feasible into a single, coordinated system of activities.

3. Equipment layout and operation sequence should be designed so as to optimize the flow of material.

4. Reduce, eliminate, or combine unnecessary handling procedures and or activities.

5. Make unit loads as large and/or flow rates as rapid as is consistent with production requirements and/or product quality.

6. Mechanize whenever possible and economical.

7. Whenever possible, choose equipment that can provide a variety of tasks and handle a variety of materials and/or products.

8. Use specialized handling methods and equipment only when absolutely necessary.

9. Whenever possible, plan for optimum utilization of handling equipment and manpower.

10. A preventive maintenance program should be planned; repairs should be made on a scheduled basis rather than waiting for a breakdown to occur.

11. Inefficient methods and/or equipment should be replaced just as quickly as more efficient methods and equipment are available and can be justified. Handling procedures and activities should keep pace with the capability and level of automation of production operations.

12. Determine the effectiveness of handling procedures and systems in terms of expense per unit or quantity handled.

REFERENCES

The Economics of Automated Warehousing, published by Control Flow Systems, Inc., Lancaster, Pa. 1970.

Industry Case Examples

CASE 3A

Programmed Handling

Automated handling provides exceptional production and scheduling flexibility as well as improved productivity. On this single assembly line at Chrysler's Missouri Truck Assembly plant shown in Figure 2, compact unitized vans and lightweight pickup trucks with separate bodies and chassis are produced simultaneously. Any combination can be handled by the coordinated system of power and free overhead conveyors and floor-mounted chain conveyors.

At a speed of ten feet per minute, up to 96 vans and 144 pickups can be assembled in one eight-hour shift.

CREDITS: Link-Belt Engineering Group, FMC Corporation.

CASE 3B

Handling for Assembly

In manufacturing plants, the in-floor towline conveyor system consolidates diverse plant activities into a smooth-flowing operation while eliminating floor clutter, controlling work backlog, and improving supervision.

Switch-carts serve as mobile workbenches for assembly, testing and inspection operations, then to carry finished product to storage or shipping areas (or simply hook a towpin directly to the chassis of a vehicle being produced on a line). The cart (or towpin) is programmed to stop or pass by only predetermined locations, (see Figure 3) depending on product requirements, where any and all sides can be oriented to an operator.

Switch-carts carry materials to spur locations convenient to various manufacturing operations, and there are unloaded either manually or automatically. Finished products are manually or automatically loaded on the empty carts and programmed for other destinations (see Figure 4).

In Figure 5 are shown several stages in the assembly and testing of electric motors where in-floor towline conveyors consolidate a multitude of plant activities into one smooth-flowing operation.

64

FIGURE 2.   Overhead handling of automotive bodies in production.

Basically, the system paces the work flow, providing controlled backlogs of work at each station, increasing productivity; it reduces the time between work stations, better controls inventory, eliminates floor clutter and many unproductive functions. The in-floor towline automatically moves products at controlled speeds through various operations.

CREDITS: SI Handling Systems, Inc.

FIGURE 3.   Programmable tow cart showing destination data.

FIGURE 4. Automatic tow cart being set for destination.

FIGURE 5. Automatic tow cart system carrying motors to next operation.

4

Automatic Control for Automation

The control function is the key to creating successful automated production systems. The more extensive and/or complicated the process, the more important it is to arrange the various process operations in the sequence most conducive to efficient product manufacture. Automating such a process not only means mechanizing many of its operations but, equally important, interlocking them via a centralized control so that the overall process will function as a single intergrated system.

The term "control system," then, refers to the means used both to automate production equipment and to integrate the operation of various equipments. Generally speaking, such a system is manual if an operator is required to start, stop, and/or adjust the process by pushing buttons or turning knobs, regardless of how many sophisticated components may be interposed between the operator and the final controlled quantity.

69

With an automatic control system, there is no manual manipulation of the process. Once the sequence and timing cycle has been initiated, the process goes through its predetermined cycle steps regardless of how many or how intricate they may be.

CONTROL FUNDAMENTALS

To correctly understand the nature of the central role taken by automatic control systems in relation to automated production, one must begin with the fundamentals. These fundamentals include: first, elements of an automatic control system; second, basic classes of automatic control systems; third, operational modes of feedback controllers.

Elements of an Automatic Control System. All control system components are basically energy storage or transfer devices. The essential elements of an automatic control system are the controlled variable, measuring element, final control element, and controller. A *controlled variable* is the measurable quantity related to the desired properties of the process or product to be controlled. Temperature, pressure, and speed are examples of controlled variables. A *measuring element* is a device − such as a thermocouple, potentiometer, etc. − used to measure the value of the controlled variable. A *final control element* or *actuator* is an apparatus, such as a motor or a valve, capable of effecting a change in the controlled variable. A *controller* is a means for operating the final actuator element in response to signals from the measuring element.

Measurement, therefore, is the key to control. Without knowing the existing states of crucial process variables, one cannot alter those states to achieve a new, more desirable process condition.

Basic Classes of Automatic Control Systems. There are two general classes of automatic control systems − open-loop control and closed-loop or feedback control. In the former class of control system, control is effected in accordance with some arbitrary reference quantity. Satisfactory action is dependent wholly on the linearity of the control mechanism or on calibration. An input signal or command is applied and amplified, and an output is obtained. The actual output may vary from the expected output to the extent that time, temperature, humidity, etc., affect the components of the system.

Closed-loop automatic control systems include the same fundamental components as open-loop systems, but have additional features. In a closed-loop control system, actual output is measured, and the signal corresponding to this measurement is fed back to the input station where it is compared either to the input or to a desired output. This kind of control is self corrective − that is, any deviation from prescribed process parameters results in an error signal that automatically activates a compensating action. There is also the closed-loop feed-forward or predictive system that is coming into use.

It is the closed-loop (feedback) control system that is utilized most often to control today's automated production process.

Web and Fluid Process Control. Automatic feedback process controllers function according to changes in the controlled variable or variables as indicated by the measuring element, and produce the counteraction necessary to maintain some desired state of the process. The method by which the controller produces the counter-action is called the mode of control.

There are four fundamental feedback controller modes used in process control:

1. Two-position control. This is the simplest form of feedback control. Sometimes called an on-off control, this mode of operation is one in which the final control element is moved to the highest (on) or lowest (off) position when the controlled variable reaches a predetermined magnitude.

2. Proportional control. The next step in controller sophistication is to vary the regulating action in proportion to the magnitude of the deviation from a reference value or set point. With this control mode, there is a continuous linear relation between error values and controller output. Proportional control eliminates the oscillatory pattern of control which is characteristic of the discontinuous on-off controller.

3. Integral control. With integral control, there is a predetermined relationship between the value of the controlled variable and the rate of motion of the final control element. The final control element is gradually moved toward either the open or closed position, depending upon whether the controlled variable is above or below predetermined magnitudes.

4. Derivative control. Derivative control action is usually used in conjunction with the proportional or integral modes. When present in a proportional or integral controller, this mode provides an even faster and more exacting correction than any of the other modes alone.

Actual process controllers are available with a single mode of control action (as in the case of on-off, proportional, or integral controllers) or in combinations of more than one mode (as with proportional plus integral, proportional plus derivative, etc.). The three controllers that are used most in process control applications are units utilizing the proportional, derivative, and proportional plus integral modes.

DISCRETE PROCESS CONTROL

In the control of discrete manufacturing processes single and combination systems can be utilized that employ various mediums to transmit control signals and/or actuate control mechanisms. The mediums used are: *mechanical, electrical* and *electronic, fluid* (hydraulic and pneumatic), and combinations thereof (such as electrohydraulic). Each medium or combination of mediums has certain advantages and limitations which makes them more or less amenable to certain kinds of applications.

The mediums most widely used in automatic control systems for discrete production processes are electrical and electronic, hydraulic, pneumatic, and various combinations of each.

Electrical and Electronic. Conventional electrical and electronic control systems utilize various arrangement of relays, stepping switches, etc., to perform specific logic functions which, in turn, result in output signals that directly or indirectly perform the necessary on, off, and/or regulating actions on control of power drive systems. Most machine control systems are at least partially electrical or electronic in nature. Where relays and their associated devices do not meet speed, space, and/or reliability requirements solid state approaches are used. Magnetic, transistor, and integrated circuit logic are all being used extensively to control the actions of the drives required to operate the automated process.

The most popular combination automatic control and drive systems that use electrical or electronic means are electromechanical, electropneumatic, and electrohydraulic systems. With electromechanical systems, electrical or electronic circuit logic is combined with electric motors and the appropriate gear trains. In electrohydraulic systems, electrical or electronic control circuitry is used in conjunction with hydraulic pumps, valves, etc., to accomplish automatic process control where motors provide the primary input. In both cases, transducers of the kind required are used to sense the process variable to be controlled.

Hydraulic. Hydraulic control systems are used most often to perform the "muscle" tasks of automatic control. Because of their ability to exert and hold large loads, these systems are found operating machine tools, forming presses, molding machines, etc.

For continuous speed control — including most forms of servo control — hydraulic actuation is convenient and reasonably cheap, provided that valve control can be used. Where high energy conversion efficiency is necessary, both hydraulic and electric servos are in use. In the hydraulic versions, variable delivery pumps must be used.

In general, hydraulic systems perform the actuating functions of machinery operation. Compared with electrical controls, hydraulics is less effective when the functions of signaling and computing are involved. In certain applications, the optimum control solution is electrohydraulic in nature — with the electrical or electronic portion performing the signaling and computing functions and the hydraulic portion performing the actuation of "work" functions.

Pneumatic. There are three basic types of air control systems: (1) pilot circuitry uses standard directional air valves in various combinations to perform control logic functions; (2) fluidics uses components that have no moving parts, and these components make use of the properties of fluids passing through them to perform basic switching actions; and (3) moving-part logic systems utilize small, modular directional valves of the plug-in diaphragm, spool-in-body, or packless spool and sleeve types.

All three of these basic pneumatic control systems share the advantages of using compressed air as a means of transmitting signals and/or actuating mechanisms. Air is readily available, clean, can be exhausted to the atmosphere without detriment, is safe, requires simple circuit designs to accomplish control functions, and is easy to maintain.

Which type of pneumatic control system is chosen for a particular application will depend on which considerations are the most crucial. The principal advantages of fluidics systems are that they have no moving parts, are small, and can be easily interfaced with other mediums. At its present stage of development, fluidics can be expensive because there are relatively few systems people who can tailor the available hardware to specific production requirements.

SENSING AND MEASURING INSTRUMENTS

Whatever can be sensed can be measured. Whatever can be measured can be controlled. There is a variety of instruments available to sense and measure process variables, although much work remains to develop the sensing and measuring devices having accuracies adequate for today's requirements (see Figure 1). The outputs of these instruments serve as the inputs for automatic controllers which, in conjunction with associated hardware, comprise an automatic control system. Though the number and variation of process measuring and control instruments are great, they are based upon the operation of a relatively few basic devices. The devices, in turn, operate on the basis of an even fewer number of fundamental physical phenomenons. The best way to categorize sensing and measuring instruments is by the process variable they measure (see Figure 1).

NUMERICAL CONTROL

Numerical control (N/C) systems have come into their own in the last few years in the area of automating discrete parts production. In concept, N/C exists

Variables Measured	Control Actions	Character of Control	Control Devices	Sensors	Actuators
Temperature	Indicate	Mechanical	Computers	Transmitters	Motors
Flow	Record	Electrical	Logic Devices	Detectors	Actuators
Electrical Characteristics	Signal	Electronic	Telemeters	Counters	Servos
Force (Strain)	Alarm	Hydraulic	Starters	Instruments	Pumps
Events	Control	Pneumatic	Controllers	Signals	Adjustable Speed Drives
Level		Nuclear	Recorders	Switches	Constant Speed Drives
Vibration		Fluidic	Relays	Gages	Compressors
Pressure (Differential)			Analyzers	Meters	Transformers
Chemical Analysis			Timers	Fluidic Devices	Clutches
Weight			Relays	Servos	Brakes
Physical Condition (Size)				Transducers	Regulators
Time					Valves
Specific Gravity					Hydraulic Cylinders
Electronic Characteristics					Pneumatic Cylinders
Position (Motion)					
Speed (Acceleration)					
Physical Analysis (Properties)					
Count					
Moisture					
Vacuum					
Chemical Conditions					
Displacement					
Pressure					

FIGURE 1. Measurement and control elements.

somewhere between the wired-in logic approach of conventional controls and the programmed instruction technique of process computers.

With N/C, machine actions are controlled by the direct insertion of data at some point. Most N/C machines get their instructions in digital form. This is accomplished by mechanically punching coded symbols, letters, and numbers on paper tape or cards; or electronically recording them on magnetic tape. The coded information performs like a "mini-program." The control "interprets" the code and directs the machine according to the sequences contained therein. Most recently the trend is toward use of interactive graphics for design of components with the computer output being a tape or direct digital transfer to the N/C machine, bypassing the drawing stage completely.

N/C has been successfully interfaced with computers in two ways. First, control computers have been used to "supervise" the operation of several N/C machine tools. Second, N/C systems have been accessed to a central computer via remote terminals. This latter direct numerical control (DNC) system allows on-line parts programming to be accomplished. What are the advantages of N/C? It:

a. Makes possible a high degree of production flexibility.
b. Produces unmatched machine accuracy and repeatability.
c. Allows high machine utilization.
d. Greatly increases productivity.
e. Results in substantial tooling savings.
f. Permits reduced lead time.
g. Requires less workpiece handling.
h. Produces parts that were once thought to be impossible.

There are certain production situations that are basically unfavorable to N/C. It is not advantageous when:

1. A manufacturer cannot economically justify purchasing and/or using an N/C machine. If your production is in large lot sizes, requires conventional tolerance conformance or repeatability, requires little change in set-up, and requires little or no operator intervention, N/C is *not* for you.

2. A manufacturer — even though the purchase and/or use of an N/C machine can be justified — cannot adjust his thinking to fully exploit (via quality control etc.) the advantages of N/C.

DIGITAL DEVICES AND COMPUTERS

All of the control systems discussed thus far have the same limitation: once they are designed and installed, any basic change in control scheme requires a corresponding change in the physical components and arrangement of the control system elements. The control scheme is "wired-in" so to speak. There is little flexibility in the respect that changes in process characteristics necessarily require a change in the logic governing them, which means a change in the physical components that produce the control logic.

Computers change all that. By interfacing a programmable master control (the computer) with conventional (electrical, hydraulic, and/or pneumatic) machine control systems, the capability is created of altering a control scheme without tampering with the hardware used to carry out that scheme. Whereas "general instructions" (i.e., how a function, machine, or sequence of operations will be actuated, etc.) "wired-in," "Specific instructions" are not. With a computerized process control system, the physical components of the process are "connected" to the on-going decisions of management. Management can alter process requirements, interrupt process operations, etc., by altering the program, i.e., the software by which the computer hardware directs subsidiary control systems. It is this concept of an overall, master control via computer that will increasingly characterize automated production processes of the future (see Figure 2).

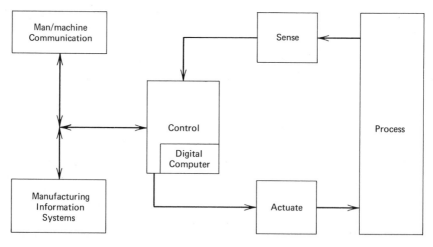

FIGURE 2.   Role of computers in process control.

Digital techniques are the basis of computer design and operation. A digital device operates in a discontinuous mode as opposed to the continuous mode of an analog device. An analog (continuous) signal is broken up (digitized) into pulse trains of discrete bits of data that can be converted into numerical quantities. The binary (base 2) number system is used. Digital techniques lend themselves to both electronic and pneumatic implementation. Arrangements of electronic or pneumatic circuit elements result in devices that can perform specific switching and/or logic functions.

The advantages of digital controls include:

1. Because such a control system regulates on a pulse-to-pulse basis (i.e., in the kilocycle per second range) corrective action is more immediate, accurate, and repeatable.

2. Digital circuits provide relative freedom from the effects of control component drift.

3. Precise operating parameters can be established over a wide range in small, precise increments.

With this background as to the nature of the digital techniques that underlie computer operation, we might ask: How does a computer work? When should you consider applying a control computer profitably to your own production process? The answers to these questions are crucial to any manager's grasp of the developing nature of manufacturing automation.

Elements of a Computer System.   A process control computer system consists of the following basic elements:

1. An arithmetic unit that multiplies, divides, adds, subtracts, changes signs, and compares numbers.

2. An instruction counter that indicates which instruction is being executed.

3. A control unit in which logic circuits interpret the instructions and direct the remainder of the computer's functions. Arithmetic and control units, together, comprise the central processing unit (CPU).

4. A memory that stores numerical data and provides them upon request from the CPU.

5. Input and output devices that permit communication with the process and/or human operators.

6. Input/output channels to enable input-output operations to proceed at the same time computations are being made.

Instructions are coded in numerical form (i.e., as digital data) and arranged into programs. A digital computer interprets and executes these instructions. If the computer system is "on-line" − operating in "real time" − then disruptions to the process are immediately fed back to the computer where they are evaluated in terms of programmed requirements and compensated for by subsequent, computer-generated changes in the process.

With direct digital control (DDC), all the diverse sub-control systems and their respective processing operations and machines can be integrated into a single integrated system which can be communicated with via software programs. This capability means that what might have been a series of related process steps can now be operated as one system. Optimization can now cover the inter-relationship of all process elements as well as each separate element.

Application Considerations. Originally, computers were applied largely to continuous processes. However, with the rapid evolution of the minicomputer, digital computers are now being applied widely in discrete parts production as well. Minis operate inspection operations, direct the functioning of one or more machine tools, and interface with numerical control equipment to provide on-line parts programming and cutting operations. The more restrictive the application, the smaller the computer that will be required to accomplish it and, correspondingly, the less programmable capacity that will be available or necessary.

Wherever the power of a computer is not required, the "programmable controller" can be used. Simpler than a computer, yet similar, it is programmed using relatively few easy instructions. The programmable controller is well adapted for such functions as sequencing, counting, timing, and other control functions for controlling machine tools, conveyors, assembly machines, conveyors, handling devices, testing machines. Having fewer moving parts than standard relay control panels they are often called "relay replacers."

Computer control offers high flexibility in operation. Changes can be made in computer programs to: (1) modify and improve control actions as indicated by operating experience; (2) change the controlling action from one process to another; and (3) change from a data logging or monitoring function to a fully automatic on-line function. In specific installations, a single computer could be used to control several unrelated processes at the same time.

What kinds of processes lend themselves to computer control? Important process characteristics include: (1) large size, (2) frequent process disturbances, (3) high complexity, (4) thorough understanding of process parameters and operation, (5) adequate instrumentation, (6) continuing importance of the process to future business goals, and (7) availability of many kinds of technical specialists.

A process control computer is not just another control system that can be integrated into existing production operations. Unless the computer is small and is being applied to a single pilot operation, the use of a process computer will necessitate reorganizing the manufacturing facility, hiring and/or consulting with technical experts, and retraining existing personnel if the installation is to be successful.

For these reasons, because of unfamiliarity with these new techniques and owing to the imperative need for assured reliability, the transition from the tried and true analog control to direct digital control via computers has been most cautious. Slowly, as systems are being evolved one by one, experience is growing and rapid gains in control knowledge are being accumulated. In the minds of those who have successfully developed DDC systems, this mode of control will be eventually accepted on an industry-wide basis. The time scale will be determined largely by economics and displacement of analog orientation among practitioners.

Computer control can in no way substitute for managerial competence. To the contrary, managerial incompetence coupled to a computer is guaranteed to create a quicker and more thorough chaos than was ever possible before. But, studied adequately and applied competently, DDC process control systems will add a new vista to production profitability.

CONCLUSIONS

The man who invented the wheel must have thought that overland travel was the last word in human transportation. Likewise, manufacturing engineers and managers are prone to look at every control advancement as the last major control development. But the productivity pressures necessitating automation will not cease, thus, progress must not halt. There is no doubt that – no matter how "way out" state-of-the-art control technology may seem today – there will be further, more "incredible" developments tomorrow.

Not only will improvements be made in the usual types of sensing devices. More accurate and more expensive measuring devices are already and will be more easily justified in the future to attain critical cost objectives. In many instances consumer demands and legal liability will justify their use.

In addition, other new components of sub-control systems and new concepts of totally-integrated control systems are already in the laboratories. For example, future computer process control systems will be decentralized so that decision makers will have access to the central computer and each other via remote terminals or even personal minicomputers. In this way, decision making will go on-line to keep pace with process automation.

For one thing, we can expect to see the emergence of what can be called total production monitoring. Under this concept, control in t' ᵃ automated plant will be complete with management information for long-rᵃ ge decision-making, a resultant byproduct of the system. Included in such systems will be some or all of such control activities as: inventory reporting and control, tool scheduling, tool stores management, shop floor control, contract status reporting, machine cycle timing, machine sequence control, trouble shooting, preventive maintenance scheduling, facilities monitoring, quality assurance, warehousing.

Adaptive control will emerge gradually as a part of the entire control picture. For example, in metalworking, tool wear and production quality will be monitored for such factors as power consumption, tool vibration, speed correction to maintain finish quality, detection of out-of-limit size, and tool change to produce optimum life and production costs.

The only way for a manufacturing company to be invulnerable to competitive pressures is to lead the competition. A vigorous development program will help to discover, develop, and apply the new control technologies that will be necessary to maintain profitability in tomorrow's industrial environment.

Those who do not strive to lead are doomed to follow. Tomorrow, more and more followers will find it hard to stay in business.

REFERENCES

Progress in Direct Digital Control, edited by T. J. Williams and F. M. Ryan, Instrument Society of America, Pittsburgh, Pa., 1969.

Production Automation and Numerical Control by William C. Leone, The Ronald Press Company, New York, 1967.

Process Instruments and Controls Handbook by Douglas M. Considine, Editor-in-chief, McGraw-Hill, Inc., New York.

Handbook of Applied Instrumentation by Douglas M. Considine, Editor-in-chief, McGraw-Hill, Inc., New York.

Encyclopedia of Instrumentation, Computing, and Control by Douglas M. Considine, Editor-in-chief, McGraw-Hill, Inc., New York.

"Minicomputers in Industry" by S. P. Jackson et al, *IEEE Transactions on Industrial Electronics and Control Instrumentation,* May 1971, vol. 18, no. 2, Institute of Electrical and Electronics Engineers, Inc., New York, p. 53.

"Control Trends," *Metalworking News,* June 21, 1971, Fairchild Publications, New York, p. 21.

"Automating the Factory," *Business Week,* June 5, 1971, McGraw-Hill, Inc., New York, p. 82.

"Guidelines for DNC," by R. E. Reed, *Automation,* May 1971, Penton Publishing Co., Cleveland, p. 61.

Industry Case Examples

CASE 4A

Automatic Assembly Systems

The Buhr Machine Tool Corporation has developed a unique cam-operated nonsynchronous assembly system for large assembly machines.

Mechanical control for automatic assembly machines is exemplified by this system built to produce rear drum brake backing plates at the gross rate of 900 pieces per hour (see Figure 3). Despite the emphasis on front disc brakes for U.S. automobiles, the increasing need for rear drum brakes has continued unabated so industry is seeking methods for improving production rates.

The unique feature of this system is the use of screw machine-type drum cams, bellcranks, and levers to power motions within the assembly stations. This system permits gradual acceleration and deceleration of moving members for smoother operation, overlapping of motions for shortest possible cycle times, and a reduction in the number of controls needed.

Cams also guarantee that the sequence timing of motions will not be altered even if the station is speeded up for increased production. To provide the safety overload protection needed with positive cam drive, each motion is provided with an overload sensing device that will stop the cam drive before any damage to the station can occur. An operator can then free the jam manually and have the station back in operation in a few minutes.

The backing plate system operates on a four-second cycle which is very fast for a system of this size. However, several stations are capable of operating even faster if needed. A speed-up is possible because each station has its own drum cam. Powered by a continuously running electric motor, the drum is driven through one revolution each cycle by alternately engaged air-operated clutch and brake units.

This machine includes automatic loading, automatic feeding of a reinforcement plate and anchor pin, welding of the reinforcement, hot upsetting of the anchor pin, hollow milling of the anchor pin and finally, sorting of good and reject parts during automatic unloading.

CREDITS: Buhr Machine Tool Corp., Subsidiary of the Bendix Corporation.

81

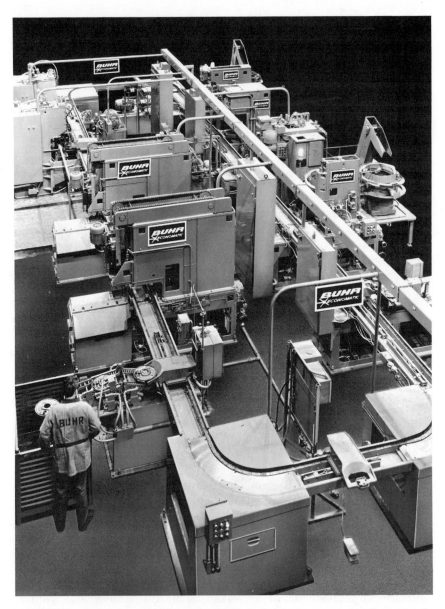

FIGURE 3. Buhr Assembly Machine. This is the new cam-operated nonsynchronous automatic assembly system developed by the Buhr Machine Tool Corporation, Ann Arbor, Michigan. This system can produce rear drum brake backing plates at the gross rate of 900 pieces per hour. (Photograph from the Bendix Corporation, Ann Arbor, Michigan.)

CASE 4B

Cement Plant Controlled by Electronics

Very few people appreciate the enormous size and complexity of a large cement manufacturing plant. The Medusa Portland Cement Company has six of these huge complexes around the country which produce tens of thousands of barrels of cement each day (see Figure 4). Contrary to the apparent simplicity of the cement process, the actual procedures involved in producing portland cement are very intricate. There are a large number of variables that have to be carefully monitored and closely controlled. The stringent requirements for such a control system can be effectively met by the versatile capabilities of solid-state electronics and the engineers at Medusa are very well aware of this fact.

Formerly, all control of process steps was done on an individual station level. Each stage of the operation (raw milling, finish milling, kiln, etc.) had its own operator. Furthermore, most of the control systems were pneumatic. "But about ten years ago," says G. MacDonald, senior electrical engineer with Medusa, "we began introducing electronic controls. Today, we buy nothing but electronic control systems. We don't buy pneumatics anymore.

MacDonald, an electrical engineer with General Motors before coming to Medusa, and another engineer handle most of the electronic design or contracting for the cement company. "In really big system installations, we call in consultants and operate as project managers and liaison for the project. We also have a plant engineer and electrical maintenance men at each cement plant."

Medusa buys a large amount of individual equipment and complete control systems, some of them costing as much as $500,000, from electronic manufacturers: "We deal in many types of electrical and electronic equipment," MacDonald explains, "motor controls, unit substations and electrical power distribution." Most of this equipment is bought when Medusa is building a new plant or expanding one of their older ones. But here too, the trend is to replace relay systems with solid-state electronics.

What is the result of using all this electronics?"Well, it's hard to calculate it in dollars and cents. When we began introducing electronics about ten years ago, we also introduced new process methods and refinements so we had to allot the credit for increased efficiency. But there is no question about having a much better and more highly efficient operation. The most obvious effect of electronics is centralized control."

In the newest of Medusa's six plants, a single operator, located in a 30 x 40 ft. room, can monitor and control every aspect of the huge plant's complex process (see Figure 5). This has been made possible through the use of the latest closed loop electronic control systems and two high speed digital computers. The large plant, which uses a single rotary kiln capable of producing 12,000 barrels of cement each day, involves about 150 variables. Control of these variables is accomplished by a number of sensing devices: thermocouples for process temperatures, magnetic flow meters for material in process flow rates, transducers and differential pressure cells, intricate level measuring devices, etc. Even nuclear devices are used which measure the density of the slurry.

Obviously, down time of the entire production line would be prohibitively expensive. According to MacDonald, the failure of a single piece of electronic equipment would not shut down the entire process. "Besides," he explains, "there is a built-in manual override in each step of the process. Of course, if all the electronic gear went at the same time, we would have a real problem on our hands. But that's not likely to happen."

As in the case of other areas of heavy industry, noise is the greatest problem for small signal semiconductor electronic equipment. "We've got drive motors as big as 5,000 hp; they can produce some pretty large inductive spikes. So the sensitive electronic units are well shielded." Aside from shielding, a special power supply system has been built for the digital computers to overcome plant generator spikes. Power from the utility is used to drive a large motor. The motor is mechanically coupled to an alternator on which is mounted a heavy flywheel. The flywheel momentum maintains the speed of the alternator in case of a momentary electrical power interruption. The well regulated output of the alternator is used for the computers. This system can actually ride through a 15-second electrical power interruption.

The significant advantages of computers and electronic control systems are making deep penetration into the cement making industry. Medusa is only one of about 50 producers of cement (Medusa ranks about 15th in size). They are all turning to electronics and solid-state control systems to solve their production problems. That means additional new markets for electronic products and also, a need for electronic engineers to design and maintain those systems.

CREDITS: Courtesy *Electronic Products* magazine, October 1970.

CASE 4C

Controlling Machines by Computer

Automation computers are designed to provide reliable automation solutions in the manufacturing environment. An example of manufacturing automation is an automatic three-axis machine tool used in the winding of coils. This and other types of numerical control machines are being used by major manufacturers to provide increased production and higher quality through greater accuracy and lower rejection rates.

The SPC-12 computer is the central control element of this three-axis machine tool (see Figure 6). The SPC-12 generates the sequence of control signals required to move the machine on its three axes as well as monitoring the

FIGURE 4. Aerial view of Medusa Portland Cement Company's Charlevoix Michigan Plant.

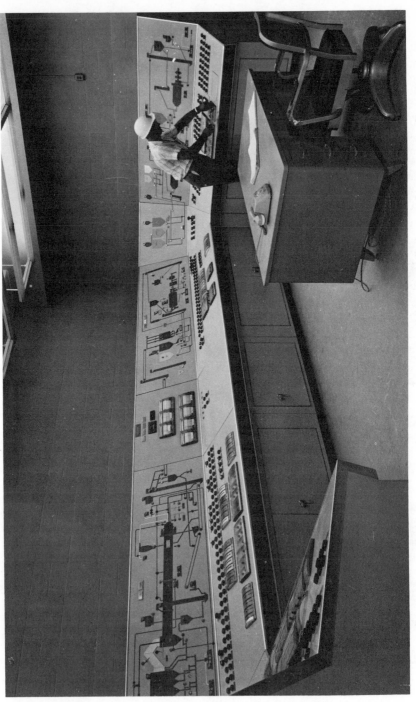

FIGURE 5. Semigraphic control cabinet for kiln and finish mill. Medusa Portland Cement Company, Cleveland, Ohio.

machine's actual position to ensure the system's accuracy. In addition, displays showing both the system's parameters and any error conditions in the machine tool are available on demand.

In the full system, one SPC-12 computer controls a number of machine tools independently. Any of the several patterns stored in memory can be selected to run on any of the machines simultaneously.

Because of the SPC-12 Automation Computer, the production rate is significantly increased because hand work is practically eliminated and the rejection rate is reduced to nearly zero. Both of these factors substantially reduce the cost of the finished product. In addition, computer programming takes all of the tedious manipulation of the adjustments out of the hands of the operator.

SPC-12 computers have been used to control traffic, test engines, set type, drill holes, make printed circuit boards, rivet airplane wings, monitor power generation systems, control nuclear reactors, switch communications systems, mix cement, count cars, manufacture electronic components, automate lab instruments, test computer peripherals, control displays, test automobiles and carburetors, move materials and perform a variety of other industrial tasks reliably, efficiently, and profitably.

CREDITS: General Automation, Inc.

CASE 4D

Automation in Grain Handling

Traveling through various cities you have noticed large complexes of cement silos. These silos are built in groups called grain houses. The filling and emptying of these storage areas is done by a maze of conveyors. Specific grain and grades of grains are stored in specific silos and it is up to an operator to direct the grain storage or delivery process.

The control panel for an ordinary storage system will consist of 250 to 300 relays, timers, etc. This type of system lends itself well to control automation in the form of the programmable controller. At Cargill, Inc., the problem consisted of tying an existing grain house into a new grain house. This meant existing conveyor systems had to be monitored to insure that they were feeding the proper conveyor in the new grain house. The selection of the correct storage silo area further complicated the sequencing of equipment.

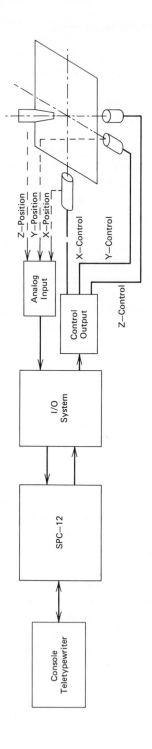

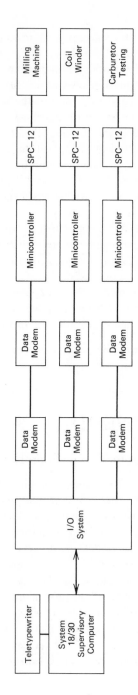

FIGURE 6. Computer control of a three-axis machine tool.

Problem areas with respect to the control are numerous when connections to existing equipment are required. The need to know when a specific existing relay or solenoid is energized is a must. But existing relays utilize all contacts and solenoids are not relay operated. The only way available for obtaining this information is to parallel new relays with the old circuits. This requires new panels, much more wiring, engineering time, panel layout, etc. An easier answer to the problem is the input to the AutoMate 33, a programmable controller (shown in Figure 7). The input circuit cards to this unit will accept 120 volt a-c. With the input across the coil or solenoid in the 120 volt a-c circuit (in conjunction with a suppressor – one wire and common per input), only engineering time is required for programming.

Inputs from the new grain house are connected to the input cards and the entire composite system is programmed at one time. No mess, no fuss. Outputs to various solenoids or starters are arranged through the use of solid state a-c switches. These a-c switches where connected with existing equipment are remotely mounted with that equipment. The a-c switches involved with the new grain house are mounted in the same cabinet (see Figure 8) as the basic AutoMate 33 equipment since it is mounted near the new starter and solenoid equipment.

The AutoMate 33 panel to handle this expansion is 86 inches high by 52 inches wide and 18 inches deep. This compares with a relay sequencing panel to do this same job of 86 inches high by 18 feet by 18 inches deep. Space saving at no premium. In addition, there is still room for the addition of some 128 future inputs or outputs in this cabinet by merely adding the appropriate input or output cards.

Creation of the wiring diagram for the unit was done by Cargill's Central Engineering Department at Minneapolis with the entire AutoMate 33 system, including keyboard-verifier (shown in Figure 9), cassette, and main cabinet at hand. Program sequencing was completed and the entire unit shipped to the job site for installation. Since wire numbers were pre-assigned for external wires at time of purchase to coincide with rack numbering, the cabinet could easily be set in place and all wires connected. Final drawings of the sequencing were made from a hardcopy runoff of the final program after start-up. This is automation from drawing board to plant operation for both greater productivity and simplicity.

CREDITS: Reliance Electric Company.

FIGURE 7. Standard production AutoMate 33 in floor-mounted 66-inch-high cabinet.

FIGURE 8. Production unit with two input/output racks at top followed by interface rack on swing frame. Processor mounted in lower swing frame.

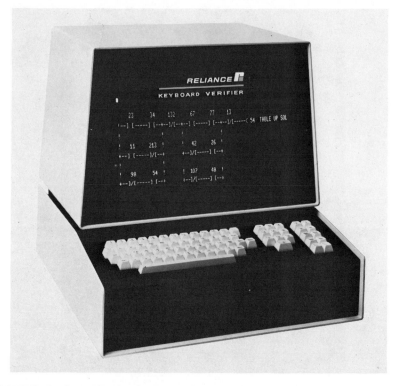

FIGURE 9. Portable keyboard/verifier showing program keys and display screen with typical relay sequence.

5

Information Systems for Automation

An imperative, often overlooked in the concentration on the machine system, is that the effective operation of a manufacturing process requires the generation, collection, and communication of pertinent *information* to the proper control points and/or decision centers for evaluation and implementation. Information control is crucial to all automated manufacturing systems. Because automated production systems function continuously at a very rapid pace, more detailed information is needed; in greater amounts; at more frequent intervals; and in more accurate forms than would be required for conventional manual production techniques.

Electronic data processing (EDP) and computer equipment constitute the hardware of information control. Here the computer is used to store, process and produce vital data on command. This area of computer use contrasts sharply with that where its capabilities are used to control equipment in the plant. Thus it must be considered as an automation tool in a different context (see Figure 1).

The emergence of EDP techniques was a prerequisite for truly advanced integration of mechanized machines and operations which automation entails.

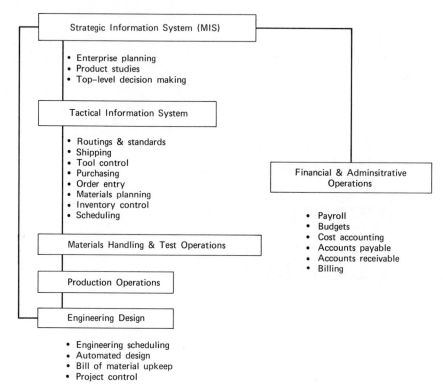

FIGURE 1. How information systems interface production and business operations.

Such integration required data concerning all the various aspects and interrelationships of a process as a whole (see Figure 2). Generating, collecting, and handling large and varied amounts of data can be a key asset in the final accomplishment of plant automation — a good example is that involved in the area of inventory control and warehousing.

TWO KINDS OF SYSTEMS

There are two basic kinds of information systems: tactical and strategic. A tactical information system deals with the day-to-day supervision and control of line activities. These activities include order processing, production scheduling and supervision, and inventory control. A strategic information system

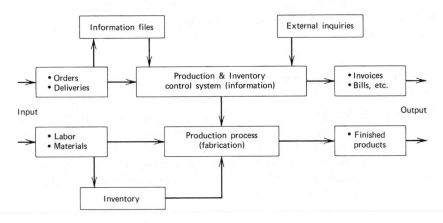

FIGURE 2. How production and inventory control system interfaces production process.

presupposes the existence of a tactical system. Output from the tactical system is accumulated and combined with financial data, market trends, business conditions, and monetary factors to provide a comprehensive input for long-range managerial decision-making. In effect, then, the strategic or management information system (MIS) is a total information system of which the tactical system is one part. Today, the tactical system is growing in use, but few as yet are going the total-systems route. Ultimately, efficient and effective strategic information systems will yield the most comprehensive and, hence, the most profitable results. Much effort is still required in order to reduce the cost of the "software" programming.

How should management proceed in the development of an effective tactical system? For one thing, it is important to "begin at the beginning" — that is, study should start with the most fundamental considerations and then proceed to higher levels of system sophistication *if and when required*. The following comments will focus on six principal areas of inquiry: First, the identification of both general and specific design objectives of any information system. Second, an explanation of the crucial role data plays in an information system, i.e., understanding what data is, how it is distinguished from information; and the necessity of maintaining its integrity throughout the system. Third, a description of the fundamental functions of an information system. Fourth, an evaluation of the significance of capturing data at its source and a description of the equipment available to do the job. Fifth, an explanation of how to apply the principles of information system design to profitable production and inventory control. Sixth, an evaluation of the advantages of creating an information division in a company to facilitate the integration of all information subsystems into a single strategic information network for maximum decision-making effectiveness.

OBJECTIVES OF SYSTEM DESIGN

Ultimately, the success of a particular information system depends on how well the basic design is implemented in day-to-day operation. But if the basic design itself is not sound, all subsequent efforts will be useless. The overall, general objective of a successful information system is to provide pertinent information at the proper places when needed, and in a form that makes subsequent action possible (see Figure 3). Specific goals are to:

1. Obtain control by exception.
2. Eliminate clerical work done by nonclerks.
3. Mechanize routine decision-making.
4. Evaluate information required for nonroutine decision-making.
5. Obtain a *minimum* of paper or report output.
6. Achieve a high degree of data processing — that is, output of information of such quality, in such form, and at such times as to require little or no reinterpretation, analysis, re-accumulation, and re-recording.
7. Obtain speed in receiving, processing, and disseminating information so as not to lessen its usefulness.
8. Provide for system expansion and refinement without fundamental reorganization and/or reprogramming.

"IN THE BEGINNING WAS DATA . . ."

Few companies today are utilizing true *information* systems. Instead, many companies employ a variety of sophisticated techniques and equipment to generate reams of useless data which supervisory and upper-management people do not have the time to review, much less evaluate. Thus, the basic purpose of the system — to control more effectively — is never realized.

The first lesson to learn about good information system design, then, is that *data is not information.* Data comprises the raw facts and figures, usually expressed in symbols, about an operation or process. Such data must be put into a form that will be understandable and meaningful to those who will use it. Thus, processing turns raw data into information. But all such information is not necessarily pertinent to the end purpose or purposes of the information system.

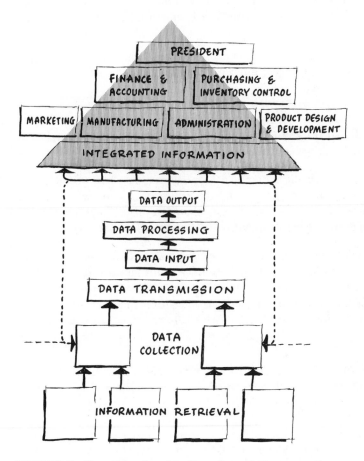

FIGURE 3. Manufacturing data flow through the enterprise.

Therefore, information relevant to the particular task or activity involved must be sifted from the overall background of processed data produced.

A further refinement is necessary. For even though all the information reaching the "sifted stage" is relevant to the tasks and/or activities in question, much of it is routine – it requires no action. Thus, only that *significant portion* of the relevant information – information that requires further evaluation and/or subsequent action – should be communicated to the appropriate control points and/or decision centers. In this way, operators and decision-makers need be concerned only with information that indicates an *exception* to usual, routine conditions.

Making sure that too much data is not generated, and that only significant information is produced, can be accomplished by satisfying the following system design steps:

1. Define the production and/or business goals, in terms of performance and profitability, that will provide a dynamic management information system that focuses on variances.

2. Determine the types and minimum amounts of information that will be needed to realize the defined goals. Avoid excessive and marginal or conflicting data. Insure against omissions.

3. Establish the nature and scope of the data base required to produce the necessary information. Isolate the critical factors on which the success of the business depends — the factors that have major impact on short-term profit results *and* long-term growth and competitive strength.

4. Design a well integrated system that will accomplish all requriements — from data generation and collection to information communication and evaluation — including a report structure that distinguishes sharply between information needed for planning and information needed for control. Be sure the reporting system identifies significant variances and why they came to be.

5. Specify equipment that will be compatible with the system design.

Generating and collecting the right amount of data for the information required is only half the problem. Equally important is the accuracy of the data. Erroneous data will yield inaccurate information. Acting on the basis of inaccurate information can be just as bad as — or worse than — not being able to act at all.

Every effort should be made to insure the accuracy of data before incorporating it into the system. One of the ways of assuring the necessary accuracy is through the technique of planned redundancy. The same data is extracted from two different sources and compared before inclusion in the data base. Redundancy will always increase the cost of the information system, but will save the time, cost, and confusion of purifying data that is discovered to be inaccurate at some later time.

It should be emphasized that any *redundancy should be planned*; the unknowing inclusion of the same data from several different sources and, perhaps, in many different forms is confusing rather than helpful, does not upgrade the data base, increases cost unnecessarily, and results in erroneous reports. Also, planned redundance is quite often used at the wrong stage. For example, it does no good to check that data is keypunched correctly if the data inputted into the keypunch is faulty.

Generating accurate data and maintaining data integrity throughout an information system requires an extensive program that encompasses four areas of activity:

1. *Education*. People who will generate input data must be made to realize the importance of proper inputs and their important responsibility in this respect. Systematic education will foster a positive attitude on the part of employees toward the system. This attitude will, in turn, make it possible for employees to cooperate to the fullest in trying to make the system work rather than trying to defeat it.

2. *Enforcement*. Management must enforce the adherence to stipulated procedures, such as observing cutoff times and providing complete data.

3. *System checks*. Self-checking, self-policing, and self-correcting features should be a part of every good system design. Where errors cannot be corrected automatically, provide for automatic error diagnosis. If this is not possible, errors can be automatically noted so that affected information will not be used unless verified.

4. *Error detection and correction*. The purposes of this function are to track down erroneous report information swiftly, and correct it at its source. Manpower for this function should be planned and budgeted for.

FUNCTIONS OF AN INFORMATION SYSTEM

When considering the development of an information system, thought must be given to the ultimate utilization of the resulting information by various managerial levels. There are five fundamental functions that are paramount in a system of this nature: data collection, data analysis, reporting, control, and communication.

1. *Data collection*. An information system will only be of value to a manager if it can provide rapid, accurate information concerning the status of his department or area of responsibility. Gathered from diverse points, data must be communicated to a focal point where determination can be made of the status of each operation. Results can be compared with desired standards. In addition, the data and resulting information can be used to determine the status of materials and manpower available to fulfill specific objectives.

2. *Data analysis*. An effective information system must not only be able to collect accurate information, but must provide for instant retrieval for analyzing data as an operation progresses. Data must be stored in such a form that

alternate actions can be determined in view of the various classifications of information in the retrieval cycle. An historical-analysis capability must be provided by permanently recording resulting information for subsequent computer processing.

3. *Reporting*. A means must be provided for supplying managers with reports concerning the status and projections of specific operations. Up-to-the-minute status reports will enable each manager to plan the operations under his direct control, as well as those operations used in a supportive role.

4. *Control*. Control is effected by an information system via a man/machine interface. Through its data collection, processing, and reporting functions, the system provides managers with real-time information that is implemented via hardware control elements.

5. *Communication*. The two prime communication functions of an information system are, first, to obtain source data from the prime areas on the production floor; and, second, to provide for centralized data accumulation and dissemination. This is usually accomplished by a combination intercommunications network and paging network.

Thus, an information system functions on two levels simultaneously: the source data collection level involving data generation, storage, analysis, and reporting; and the managerial control level involving data analysis, information communication, decision-making, and reporting for subsequent control. The more mechanized the collection, analysis, and reporting functions, the fewer clerical tasks that will be required of the manager. He can spend more time analyzing progress in relation to particular objectives, that is, he can spend more of his time being a manager.

CAPTURE DATA AT THE SOURCE

Automated data collection and accumulation is the heart of an information system. The basic concept of data collection and accumulation is the monitoring of men, machines, and materials so as to attain the optimum use of each. To develop such a capability, it is necessary to study and satisfy the needs and requirement of three main areas of a business: management, manufacturing, and finance.

It is important to capture data as close to the source as possible. All other factors remaining equal, less opportunity for error will result in fewer actual errors. Recording data at the source which generates it will insure the most accurate data possible. Source data must be recorded accurately and rapidly, preferably in a machine-oriented language for subsequent handling and processing.

There are two basic classifications of data collection and accumulation hardware: nonmonitoring and monitoring types. Nonmonitoring equipment is not capable of immediate counting or checking operations. It consists of collector units or input stations, and accumulator units or receiving stations. A collector station usually consists of keyboard, card-reader, and badge-reader modules. These modules can be interconnected to form any number of different system configurations.

Collector units or stations are placed at points near the sources of data to be collected. Badge readers may be used for recording attendance only or for both man-to-job reporting and attendance recording. They are usually placed near the collector units. Data accumulators are placed in a central location such as a tabulating room or a data processing center. Collectors are usually designed to accept and read one or more prepunched tabulating cards, and have provisions for entering data manually by means of dials, buttons, or levers. Data that enters a collector is read and transmitted to an accumulator.

Accumulators are usually designed to receive data from one to fifty collector units or stations. A scanning mechanism enables each collector to "call for time" on the transmitting circuit, and have its messages recorded at the accumulator. All messages received are checked for errors. If no errors are detected, the messages, which originate at process sensors or collector stations, are recorded. Each message is dated and further identified for future reference.

Monitoring-type data collection and accumulation equipment is specifically designed for use in the control of volume-production equipment, the output of which can be measured or counted. This equipment can also record productive and idle man-hours according to employee or job. Normally, monitoring equipment is mounted directly on production machines for counting and/or measuring each unit of work performed. Sensing devices generate pulses each time the machine is activated and/or cycled. The pulse is relayed by cable to a central control room where it is recorded. Data can be transcribed into machine-readable form and forwarded to the data-processing department, either manually or automatically.

Most monitoring systems utilize a telephone circuit as a communication link between each operator station and the central control operator. Other additions are possible to make the system more elaborate and/or automatic. The sum total capability of monitoring-type source data systems is to constantly monitor operations so management decisions can be made on the basis of minute-by-minute operating facts. In addition to its monitoring feature, all data captured by monitoring-type equipment can be collected, retained, and used in any of the ways data is used by nonmonitoring equipment.

PRODUCTION AND INVENTORY CONTROL

Application of the basic principles of information system design to production and inventory control can greatly increase the effectiveness of production operations. The principal objective of a production and inventory control system is the optimum utilization of resources at each workstation. Accomplishing this objective requires the performing of three basic types of functions: principal, supporting, and control functions. The *principal* functions of such a system include receipt of orders; breakdown of orders into jobs, and jobs into tasks; material and parts requirements; production scheduling; production control; and routing of finished work. *Supporting* functions are product specification, stock control, parts inventory, and purchase requisition. *Control* functions include keeping track of order status, work loads, and support functions. The work flow associated with these functions is based upon job-lot operation. Contraints on the functioning of the production and inventory control system are imposed by policies concerning the acceptance of delivery dates, changes in orders, and the processing of high-priority orders.

Work flow is supported by a number of information files. Information pertinent to the various functions of the system is stored in specific files where it can be retrieved easily when needed. If the system is computerized, storage is accomplished via magnetic tape. The size and structure of the information files will depend upon the nature, scope, and complexity of the production and business activities involved.

The basic operating elements of a computerized production and inventory control system are:

1. Central computer processor consisting of internal memory, program control, and an arithmetic and logic unit.

2. Source data collection and accumulation system interfaced with the computer via local and remote data terminals.

3. Auxiliary computer memory equipment.

Physical dispersion and cost of such a system will vary considerably between applications. Physical dispersion — the distances between the locations where the data terminals interact with personnel using the system — will be governed by the physical dispersion of the organization it serves. Increased dispersion implies increased data terminal costs and increased systems communication costs. Typically, the central computer processor, auxiliary memory, and main communication channel account for three-fifths of the total system cost, with interface data terminals making up the other two-fifths.

Some of the tasks performed by a computerized production and inventory control system include:

a. Daily posting of finished goods inventory, by location and product.

b. Automatic searching of inventory for products that are not available where needed, and the referral of orders to the appropriate stocking locations.

c. Finished goods inventory control, including such tasks as calculating lot size and reorder points, entering replenishment orders, and allocating production to warehouses.

d. Pricing and invoicing.

e. Processing accounts records, customer statements, and credit letters.

f. Determination of sales credit.

g. Calculation of salesman's commissions.

h. Preparation of territory control reports by industry, customer, and product group.

i. Preparation of monthly sales accounting statements.

Of course, a particular production and inventory control system need not be computerized, at least not right away. The most effective system will be one that is planned thoroughly and introduced in a step-by-step manner. As production operations become more mechanized and integrated, further refinements can be made in the information system. The end goal is to have a computer accomplishing all of the routine production, information, and decision making tasks in a company.

What steps should be followed so as to progressively create such a system? Here are some guidelines:

1. Begin by studying the information needs of a production and inventory control system. Break the system design down into subsystems so as to facilitate implementation.

2. Develop a data base consisting of part numbers, order numbers, material quantities, standard hours worked on specific tasks, operation numbers, workstation numbers, etc. Be sure to standardize the terminology used to structure the data base.

3. In the same way that the master files are developed gradually from information that already exists, the control system itself can be started by mechanizing one machine group, or function at a time; and then expanding the process until the entire control and planning loops are covered. For example, equip only one area of the fabrication department with a control center and several workstation data terminals. The data needs of that area with respect to the rest of the fabrication department and other production areas should be carefully studied. At the same time, an improvement in the total-system data base should be effected by requesting that more detailed information be given

by other areas to the inventory file for all parts produced in the selected fabrication area.

4. Gradual introduction of a production and inventory control system may mean that the collected data will not be changed to tape or punched-card form immediately. Also, gaps may be left in the inventory file for parts that are not produced by the pilot area.

5. After some experience, the controller of the control center should eventually issue a tentative manufacturing schedule for the pilot area – a schedule with a new format, more details than usual, and covering as much as ten days in advance.

6. Knowledge gained from the pilot area can be used to extend the system to other key areas in the fabrication department.

7. At this stage, thought should be given to making data files computer compatible if a computerized system is envisioned. Plans should be made for organizing the various control areas in such a way as to facilitate their integration via computer hardware.

8. The final stage is the specifying, ordering, installing, and debugging of computer system hardware.

There are many pitfalls to watch for when developing an automated production and inventory control system. Some of the reasons efforts to create such systems fail to show up in the profit ledger include any or all of the following:

a. Overly ambitious schedule for completing the system.
b. Moving too fast initially on too many different system projects.
c. Planning was not sufficiently oriented toward results.
d. Problems were not thoroughly defined.
e. Education was ineffective in overcoming employee's resistance to change.
f. Inadequate teamwork.
g. Insufficient attention to internal systems controls.
h. Reluctance to put confidence in computer operation and output.
i. Inadequate investment and/or justification programs.
j. Total cost of the full program was not fully anticipated.

CONCLUSIONS

Needless to say, improvement in the control of production and/or business operations saves money. Money should not be wasted on trying to achieve a system sophistication that is either not necessary or not possible with the talent and resources available. Efforts should be restricted to accomplishing what is necessary as efficiently and expertly as possible. Unless a tactical system is tested for its productivity improvement on a return on investment basis and found worthy, it should be abandoned. There is no magic. The key is profit or perish.

REFERENCES

"Communication or Chaos?" by Dale B. Baker, *Science,* August 21, 1970. p. 739.

Computers and Common Sense by Mortimer Taube, Columbia University Press, New York, 1961.

A Manager's Guide to Computer Processing by Roger L. Sisson and Richard G. Canning, John Wiley & Sons, Inc., 1967.

Computers, Office Machines and the New Information Technology by Carl Heyel, the Macmillan Co., 1969.

"Information Systems and Individual Freedom," an address by H. I. Rommes, Chairman of the Board, A.T.&T. before the National Industrial Conference Board, November 30, 1967.

"The Perils of Data Systems," *Business Week,* June 5, 1971, McGraw-Hill, Inc., New York, p. 62.

"Computerized Graphics – Present and Future" by Robert A DiCuricio, *Automation,* May 1971, Penton Publishing Co., Cleveland, p. 48.

"Line-Sharing Systems for Plant Monitoring and Control" by Richard L. Aronson, *Control Engineering,* January 1971, p. 57.

Industry Case Examples

CASE 5A

Chase Brass Logistics Management System

An expanded computer-based logistics management system designed to monitor all manufacturing functions has been implemented at Chase Brass & Copper Company's brass rod plant at Montpelier, Ohio. Plant management has instantaneous access to production data via a PDP-10 time-sharing computer system.

The computer, which replaces an older DEC computer used primarily for production scheduling, is interfaced to virtually all manufacturing and materials handling points in the 116,000 square foot plant built five years ago. According to Dick Ryan, plant manager, this Chase facility was designed with a computer in mind.

"The functions of the original computer system were primarily processing of orders and scheduling shipments. As we developed more expertise in using the computer we decided to install the larger computer to not only assume this function, but to provide us with a means of collecting all pertinent production information which we could use to achieve greater operating efficiencies, improved customer satisfaction, and on-line control of the casting shop."

In basic operation, a clerk enters, via a teletypewriter, incoming orders, for which the computer provides either a production schedule, based on stored historical data, or a direct shipping notice if the desired material is in inventory.

Reporting functions are kept at a minimum. "We don't feel we need reams of computer print-out on a regular basis," notes Ryan. "The computer generates a daily, one-page production report; all other information is provided on a demand basis."

All interaction with the system, including file updating, is on an on-line basis via teletypewriters or wired directly to data collection transducers. Ryan noted that "there has never been a punched card in this installation and we do not anticipate seeing any in the foreseeable future."

There are some two dozen stations on the production floor which the computer monitors. These include, the truck scale (incoming raw material), holding and melting furnaces, billet saw and inventory, extrusion press, finishing machines, and finished product scale. Information is converted from analog-to-digital format, reduced to usable form, and stored by the computer for a reasonable period of time.

In addition to its on-line monitoring functions, the computer provides an in-house time-sharing facility for plant personnel. "We have about fifteen Teletypes which are used by administrative and plant personnel for routine inquiries or for research and engineering functions."

The PDP-10 system at the Chase site consists of 32K or 36-bit word memory, one-half million word disk, a 20-channel data acquisition and A/D conversion system, special interfacing to various production machines and a line printer. A CalComp plotter is also interfaced to the PDP-10.

CREDITS: Digital Equipment Corporation.

CASE 5B

Ford Computers — Tools for the Future

Ford Motor Company entered the Age of Computers in 1955. Today, the number of computers at Ford has grown to well over two-hundred. The machines themselves have increased in power and versatility by several orders of magnitude, as components have evolved from tubes to transistors to integrated circuitry. Speed and reliability have increased enormously. The scope of computer applications has extended into virtually every aspect of company operations — including research, product development, marketing and distribution, material control and transportation, manufacturing operations, and management control systems.

The Challenge.   At Ford, the computer has become a productive tool of such size and importance that its own operating efficiency and the effectiveness with which its potential is exploited are now among the most exciting competitive opportunities. Efficiency of the computer has grown not only through improvement in hardware and software, but also through the integration of information files at appropriate corporate, divisional, and plant levels so as to minimize duplication and to maximize accessibility of information. Overall effectiveness of computer systems at Ford is measured not only in terms of economy of operation, but also in terms of responsiveness to management needs and to profit opportunities.

The Systems Analysis Approach. As computer systems have become more complex, it has become increasingly necessary to take account of the interactions among different functional systems and organizational activities, as well as environmental factors. This "Systems Analysis Approach" has been facilitated at Ford through the use of mathematical models and other techniques that cut across traditional boundaries. A key factor in this approach is the recognition that no system, organization component, or group of people can be viewed in isolation from the other company activities or from the environment of customers, public interest, and governmental policies that properly influence the operation of present-day companies.

Ford Communications Headquarters. The new Communications and Data Processing Center in Dearborn is symbolic of the increasingly important role played by computers at Ford. In addition to providing office space for Ford's credit and insurance subsidiaries, the new 300,000-square-foot building serves as headquarters for Ford's worldwide communications and data processing operations. The computer center is one of the most modern of its kind in private industry. The new center is an example of the emphasis which Ford places on the development of integrated data processing systems for the company's domestic activities.

Teletype Traffic. All teletype traffic is automatically controlled by the PCP 150 Philco Communications Processor, the hub of the company's international network of teletype communications (see Figure 4). The electronic processor receives and transmits messages and reports, shipping information, employee data, and administrative communications at the rate of 25,000 messages a day. The system services more than 250 terminals located at domestic and foreign divisional offices, parts depots, and assembly, manufacturing, and parts plants.

This centralized communication switching facility is capable of accepting message input from many sources at various speeds simultaneously. It directs the message flow to the proper destination without interruption.

Data Communication. Another large-scale computer system known as COMBAT — "computer-based telecommunications" — is employed for centrally collecting, editing, and disseminating data. The system receives and sends information on-line between the computer center in Dearborn and about seventy-five smaller computer systems and terminals at various Ford offices in North America. For some applications, the field terminals receive information unattended during the night, so that reports are available to field operating personnel at the beginning of their working day.

CREDITS: Ford Motor Company.

FIGURE 4. Ford's PCP 150 Philco Communications Processor. (Photograph from the Ford Motor Company News Department, Dearborn, Michigan.)

CASE 5C

How a Man Can Pick 600 Loads While Doing Something Else!

It's remarkable how much one man can do in the new warehouse at Vick Manufacturing Division, Richardson-Merrell Inc., Hatboro, Pennsylvania. The operator of this automated warehouse can pick as many as 600 pallet loads a day, while — at the same time — he's putting loads away in 65-foot-high racks! He can even pick pallet loads on a first-in-first-out basis, and deliver them to the shipping dock, while he's tending to other operations!

Such high productivity is made possible by a computerized storage-and-delivery system, using stackers and conveyors (see Figure 5). This system moves pallet loads of finished goods and production materials from the plant into the warehouse's stacker racks, and from the racks to the shipping dock, or back to the plant.

The storage capacity of the system is 10,000 pallet loads: up to 2,500 pounds and 60 inches high, on standard 48 x 40-inch two-way pallets. These loads can be stored 10-high in 10 long rows of single-pallet-depth racks. And this capacity can be increased to 14,000 loads by adding two more stackers, four more rows of racks, and related conveyors.

This storage-and-delivery system (Hartman Engineering, division of Hartman Fabricators, Inc.) includes five floor-supported stackers, each in one aisle between two 397-foot-long rows of racks. It also includes separate input and output conveyor networks, with an input station and an output station for each stacker.

At present, lift trucks bring pallet loads from production to feed the input-conveyor system. The plan for the future is to convey sealed cases to semi-automatic palletizers, which will feed pallet loads to the input system.

According to the project manager, Vick's industrial engineer, David J. Hower, this automated-warehouse system, with its expansion capability, can cope with the expected increase in shipping volume during the next five years. And it will also provide ample storage capacity for a variety of chemicals and packaging materials to serve production.

Stored in the warehouse are Vick Manufacturing Division and Clark-Cleveland Division products, in as many as 150 package-configurations. Generally, these products are shipped in mixed truckloads, to numerous distribution warehouses.

The warehouse-input averages about 250 pallet loads a day. Peak (seasonal) output is as much as 500 a day.

Automated Warehousing Pays Its Way: Vick's automated-warehouse system, in a building 425-feet long by 110-feet wide, provides substantial benefits.

Engineer Hower reports, "we needed a new warehouse to release floor space for increased production and to eliminate off-site storage in rental space. And we found that this 66-foot-high building, for an automated warehouse, could give us the cube we needed in about one-fourth the area of a conventional building, around 18-feet high. Yet we found that it would only cost 50 percent more per square foot!

"We also found that the additional investment for an automated warehouse should be paid off in a little more than a year!

"We can operate with one man in the warehouse and two at the dock — less than half the work force for a conventional powered-floor system.

"One great advantage of the automated warehouse is its speed. Within half an hour we can get enough pallet loads to the dock to load a truck!

"We also have tight control over an inventory that includes pallet loads waiting for laboratory release. For example, it's easy to select and pick stock on a first-in-first-out basis. And we have accurate stockkeeping by the operator.

"Of course, we could have given our warehouse computer some clerical functions, or linked it to a central computer. But, with our present methods, the warehouse operator has enough time for essential recordkeeping. So we're using the computer as a control device, just to run the storage-and-delivery system."

According to Hower, the control system, centered in a low-cost, general-purpose computer enables one man to operate the warehouse with ease (see Figure 6). In brief, the computer can run the warehouse system after accepting a series of input or output commands: up to 64 output commands, and 10 input commands — the latter only limited by the input-conveyor space.

Each command specifies the source and destination of a load — on conveyors and in racks. It includes an understandable 7-digit code, on a "storage-address" card fed to a reader. This code specifies one horizontal and vertical position in a row of racks.

The computer's memory capability is a time-saver for the operator. For example, in just a few minutes he can feed the computer a long series of output commands — enough for several truck shipments. Then, while the computerized system retrieves loads from storage, and delivers them to the dock, he can tend to other operations.

He thus has enough free time to (1) inspect and identify incoming loads, (2) dispatch accepted loads into storage, (3) maintain inventory and space files, (4) plan the next series of output commands, and (5) monitor the system.

Operating with a Computer. The input-conveyor system automatically brings loads to the control room in Figure 6. On the way, they pass through two load-sizing stations: for width, and for length and height. Any oversize loads are detected by photoelectric systems and then shunted to a reject conveyor.

Generally, the operator "programs" each accepted load as it arrives. First, he checks each load against its packing slip. Then he selects an appropriate "storage-address" card from an "empty-location" file: available storage locations, by stacker-aisle. This is essentially a random selection of a specific rack location for the load. But the operator does try to "scatter" the inventory, so that each stock item will be available in several aisles.

Next, the operator copies the card's 7-digit storage-address code onto the multi-copy packing slip on the load. Also, he removes two copies, and puts one into the envelope-section to the storage-address card. He then puts the address card in the reader and dispatches the load into storage.

Later, he mails the address card, with its packing slip, to the analytical laboratory. The second copy of the packing slip goes into a file in another office. This is a control point: where address cards of approved materials are returned from the laboratory to be matched with outstanding slips. The "approved" address cards then go back to the warehouse operator. He files them by product and by "age" in his "approved inventory" file.

The programming of an input to storage can be completed in seconds. The operator merely puts the selected address-card in the reader, and pushes a "store" button on the console. Normally, an "entered" (in computer) indicator will light up, and then the operator can immediately remove the card.

Other indicators will light up, if there are operating problems, such as: (1) load not in position to go, or (2) stacker-input capacity already assigned, with two loads already on their way to the same stacker, or (3) programming-capacity already assigned, with the commands already in the computer's memory.

The inbound load automatically goes to an elevating conveyor. This lifts it about 10 ft to a cross-travel conveyor, where a transfer shunts it to the input station for a selected stacker. Here the load is positioned for the stacker to put away, at the selected address.

Warehouse-outputs to the dock are initiated by bill-of-lading lists, showing quantities of items by truckload. The operator selects appropriate storage address cards from his approved-inventory file, on a first-in first-out basis, to satisfy shipping requirements.

To program an output, the operator inserts a card in the reader, turns a switch to "dock" (for finished goods) or to "plant" (for production materials) and presses the "retrieve" button. Normally, the "entered" indicator lights up, so he immediately removes the card. Thus he can program one output after another in intervals of a few seconds, to pick truckload quantities.

The stackers deposit the outgoing loads at the low-level output stations (See Figure 6). The output conveyor system ends at the dock in a 50-foot section of indexing conveyor, used for temporary accumulation of as many as 10 pallet loads.

HOW THE COMPUTER SPEEDS INPUT-OUTPUT

The control system (Figure 5) centered in a general-purpose DEC PDP-8 computer, can achieve high-volume input-output, despite complexities in flow. Every four minutes, each stacker can move a load in and a load out!

The computer makes this possible, through simultaneous operation of a variety of equipment in the storage-and-delivery system. For example, all five stackers can be putting away or retrieving loads, while, at the same time, input and output conveyor systems move a stream of loads to and from the stackers' input and output stations.

Effective equipment utilization is achieved by using the computer's memory for the control application.

In brief, this computer, in the warehouse operator's control room, accepts and remembers a sequence of commands, quickly entered with the card reader. Then the computer issues appropriate instructions to its subsystem controls, and receives feedback on completions or malfunctions.

To run the storage system, the computer is linked to logic circuitry in the interface. And this, in turn, is linked to the logic circuitry in the stackers' control panels.

To run the conveyor system, the computer is linked to a DEC PDP-14 programmable controller. This device has a hard-wired read-only memory, with changeable memory modules used for logic functions. This is, in turn, connected to the contact panels that operate the conveyor equipment.

The feedback from subsystem controls to the PDP-8 triggers a print-out of completions or problem conditions, at the teletypewriter.

Thus, the computer can issue new instructions, or the operator can quickly correct malfunctions, such as misaligned loads. And the operator always knows the status of the workload.

How the Conveyors Are Controlled. The programmable controller works as a slave to the computer to control the input and output conveyor systems.

For example, the controller signals the computer that an incoming load, in the control room area, is in position to move into storage. Then the computer signals the controller to move the load to a given stacker's input station.

This instruction from the computer is based on the operator's "programmed input". The programming of the operation — a "retrieve" or "store" — immediately indicates the proper address for the first function: a pickup from the storage address or from the stacker's input station. The second function then becomes a deposit, at the stacker's output station or at the storage address.

The stacker's circuitry receives the proper signals to perform these functions, in the required sequence at the right locations. Voltage signals from the interface specify horizontal and vertical position, left or right side of the aisle, and permit the pickup or deposit of a load.

Signals from the stacker to the interface report the stacker's completed functions.

An interesting feature of the stacker's logic circuitry is its system for relatively high-speed vertical and horizontal positioning. This applies a voltage comparator to seek an address.

In brief, reference voltages are provided by wipers that contact segmented conductor bars, with resistors between segments. Thus, these reference voltages change with position: increased two volts for every storage level, and one volt for every position down the aisle. And these changing voltages are being continuously compared with fixed voltage signals from the interface.

When the "horizontal" voltages are nearly equal, the horizontal drive motor is switched to low speed, so the stacker can be stopped in time. And both the horizontal and vertical motors are automatically switched from "coarse" to "fine control."

Fine positioning, vertically and horizontally, is accomplished by closed servo loops. In short, magnetic sensing heads that follow special tracks produce voltages that tend to drive the motors toward the exact "address positions" on the tracks.

So that the stacker will zero-in on each desired position, a tachometer signal also is used. The tachometer geared to the drive, produces a voltage proportional to speed but opposing the voltage of the sensing-head signal. Thus, the stacker slows down gradually and stops within 1/4 in. of the desired position – automatically.

THE COMPUTERIZED SYSTEM TAKES CHARGE

In Figure 6, the computerized system takes charge, controlling the flow of incoming and outgoing loads by simultaneous operation of five stackers and their input and output conveyor systems. Operator gives computer input or output commands by inserting coded address cards in reader and pushing "store" or "retrieve" button. To run the storage-and-delivery system, the computer issues instructions to the programmable controller that runs the conveyors and the interface that runs the stackers, and receives feedback on completions or malfunctions. The programmable controller, utilizing a read-only memory for logic functions, tracks every load in the conveyor system, through feedback from limit switches and photoelectric systems, and operates one section of conveyor after another, through its connection to contactor panels. The logic circuitry in the interface is in two-way communication with the logic circuitry in stackers' control panels. It issues signals to initiate each pickup or deposit function and receives reports on completed functions.

CREDITS: Courtesy *Modern Materials Handling*, July 1971.

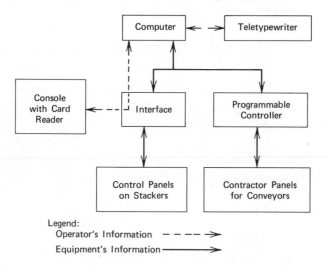

FIGURE 5. Computerized storage-and-delivery diagram for an automated drug warehouse.

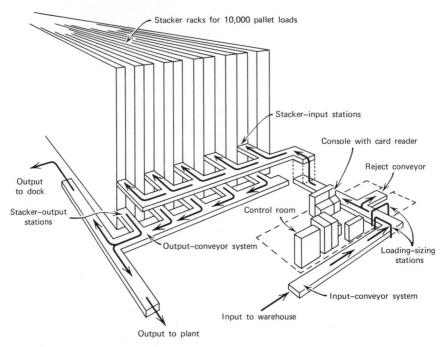

FIGURE 6. One man operates entire system by feeding cards into reader at central console. Computerized control system can remember up to 74 input and output commands and run conveyors and stackers for the operator — controlling the incoming and outgoing flow of numerous pallet loads at once.

CASE 5D

Total Manufacturing Control System

The total manufacturing control system (TMCS) is an integrated, real-time computer based system being developed by Lockheed to control an entire manufacturing facility. Beginning with engineering concept design and analysis, the system extends through all operational functions including the acceptance of finished hardware. It offers tremendous potential for reducing lead time and manufacturing costs while at the same time improving product reliability.

Evolved from today's numerical control, the TMCS is modular in design, flexible in size, and self-monitoring for dependable operation. It provides operational displays and status reports tailored to the needs of multiple levels of management. The system is designed and is being implemented with the aerospace industry in mind.

The 1970's will be a decade when computer technology will have as much impact on manufacturing — in the broadest sense of the word — as it has had on engineering, business, and financial management in the 1960s. While each new computer technique provides new capabilities, the total payoff can only be realized by integrating these techniques into a system such as is shown in the TMCS System Schematic diagram in Figure 7. The elements of such a system can be implemented in any logical order as the needs and payoffs dictate and as company resources are made available.

The principle subsystems of the TMCS are computer graphics, direct numerical control, adaptive control, in-process quality control, and real-time management control.

Computer Graphics. The computer graphics system consists of an electronic console containing a cathode-ray tube (CRT), a light pen, and a keyboard. The keyboard enables the operator to converse with the computer. The combination of the CRT and the light pen enables the operator to put information into the computer graphically.

The two methods of communication may be used singly or in combination, whichever is more appropriate. Either way, the full power of the computer and its software is at the command of the operator at the graphic console. As information is entered, it is converted to digital form and stored for future use. A number of computer graphic consoles can be simultaneously serviced by a single computer.

The System Schematic shows computer graphics consoles in both the product engineering and the manufacturing engineering complex. The consoles in product engineering are for engineering analysis, design, and release control. When necessary, the power of the APT computer can be used to support the graphics computer. Engineering designs are normally viewed at console CRT's, although provision is made to produce permanent copies of the display when necessary. If a scale drawing is required, it will be produced on the plotter.

The graphics designs are transmitted in digital form within and beyond the plant by wire rather than by messenger, as is currently done with blueprints. After engineering designs are completed, checked, and released, they are available for call-up in the manufacturing engineering and control center. There the engineering designs are worked from directly for part programming, for tool design, and for quality control purposes. This is a big plus for TMCS because, with the blueprint, part geometry had to be completely redefined in part-programming language, a costly and error-prone method.

Conversational Programming. In the 1970s, the computer graphics console and the conversational programming terminal are expected to become a single unit. When this happens, the conversational programming, as a subsystem of direct numerical control (DNC), should be capable of preparing all new DNC programs and modifying existing programs, regardless of program complexity.

Some DNC systems currently have a conversational programming capability. The extent of this capability varies according to the programming language used. Conversational programming at the machine tool is most valuable when it is necessary to correct mistakes in part geometry, eliminate wasted motions during machining, or optimize feeds and speeds. Having the program under the control of the computer rather than on tape makes all of this possible (see Figure 8).

Conversational programming speeds program proofing from weeks to hours or even minutes. It eliminates the repeating of job setups on the machine tool that occur when the program is found to be incorrect and another job has to be scheduled for the machine while the program is corrected. Conversational programming not only makes better use of the time of the programmer, operator, and machine tool, but it also imparts reliability to the entire operation of production scheduling and control.

To complete the job of program proofing accurately and quickly, on-the-machine part measurement is required (see Figure 9), often it is necessary to compensate for cutter, workpiece, or holding fixture deflection — or an occasional program error — each of which can only be detected by accurate measurement of the part. The on-the-machine probe measuring system (see "In-Process Quality Control" below) provides this measuring capability.

Direct Numerical Control. With DNC, the program is stored in a disc file provided for the computer rather than on tape. The tape reader is eliminated along with the significant cost of operating and maintaining a tape system. The large cabinets of electronic gear used for processing the program no longer occupy valuable manufacturing space.

These items that were previously required for each machine tool are replaced by a single computer which controls a group of a dozen or more machine tools directly via cable connections. DNC not only offers higher operational reliability, but it also opens the door to many new and promising manufacturing innovations, some of which are subsystems of TMCS.

The cost of DNC can be less than the cost of the hard-wired control equipment required for conventional NC. The cost of tape, tape preparation, maintenance of the tape reader, and the hard-wired electronic package is eliminated. These items are replaced by the computer, which has higher reliability and a lower maintenance cost record. These are some of the immediate payoffs provided by DNC. While they are significant, much greater long-range payoffs are possible because of the additional capabilities made possible by DNC.

Adaptive Control. The purpose of adaptive control, as applied to the rate of machining, is to optimize speeds and feeds so that the machine tool is always being used at full capacity, consistent with part specifications and good machining practice. To accomplish this, the machine tool needs sensors at its spindle to sense such parameters as spindle deflection, vibration, and torque. The sensor signals are compared to preselected values, as determined by such factors as workpiece machinability, cutter size, cutter material, surface finish required, and the rigidity of the part and the work-holding fixture. The system adjusts the feed and speed to the maximum rates that will not damage the cutter, stall the machine, produce too rough a finish, or distort the part.

In-Process Quality Control.   The in-process quality control subsystem increases the reliability of direct numerical control to the extent that scrap is all but eliminated. Because of this level of reliability, effort devoted to final inspection can be reduced to a fraction of the present level. The in-process quality control subsystem monitors the NC system for equipment and operator error. If an error occurs, the machine tool axes movements are automatically inhibited and the system provides a diagnostic signal isolating the problem.

In-process quality control consists of three subsystems, an axes-programmed path monitor, a cutter monitor, and an on-the-machine probe measuring system. Each of these subsystems can be operated independently — or in combination as an in-process quality control system — with either DNC or NC.

Real-Time Management Control.   The direct coupling of computers to graphics consoles and machine tools provides an important link in the real-time data input source for management control (see Figure 10). Data retrieved from these locations is independent of human interpretation and therefore more accurate. Computer based subsystems in support organizations also feed real-time data to the TMCS management control computer. TMCS, in turn, supplies exception and status reports to all organizations of the facility.

The role of TMCS real-time control is primarily to assist each basic function within the manufacturing facility — for example, machine tool control, cost control, and schedule control. A secondary responsibility of TMCS is to provide displays and status reports tailored to the requirements of appropriate levels of management.

In addition to the chip-making activities discussed, most TMCS techniques will apply to other areas of manufacturing such as electronics, welding, sheet metal and plastics. Singly or collectively, the subsystems of TMCS are equally capable of reducing lead time and costs while improving product reliability in all of these areas. How well this potential is realized depends upon how well managements manage the system.

CREDITS: Russell M. McKay, Sr., Research Specialist, Manufacturing Research Organization, Lockheed Missiles & Space Company, Sunnyvale, California, "Total Manufacturing Control System," presented at the 35th Annual Machine Tool Electrification Forum, sponsored by the Westinghouse Electric Corporation, Pittsburgh, Pa., May 25 and 26, 1971.

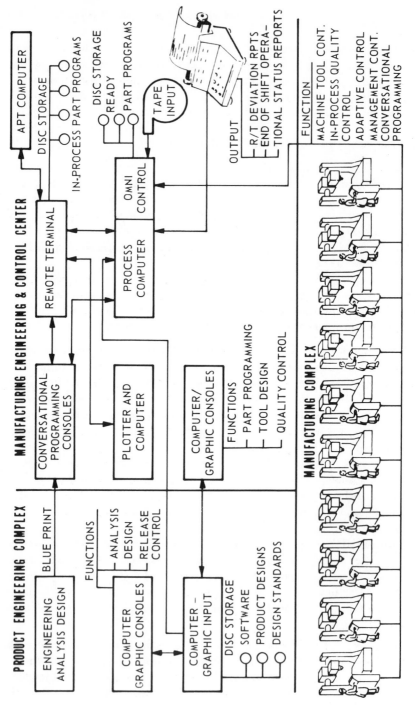

FIGURE 7. System schematic, TMCS.

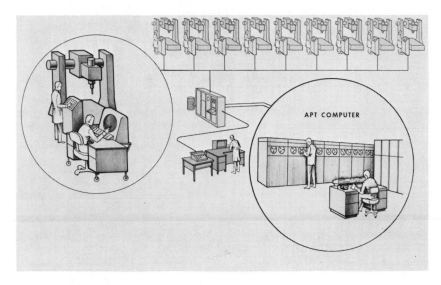

FIGURE 8.   Conversational part programming at machine tool, TMCS.

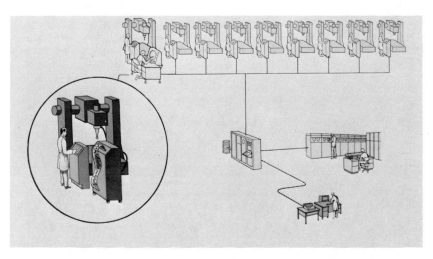

FIGURE 9.   ON-NC-Machine inspection subsystem, TMCS.

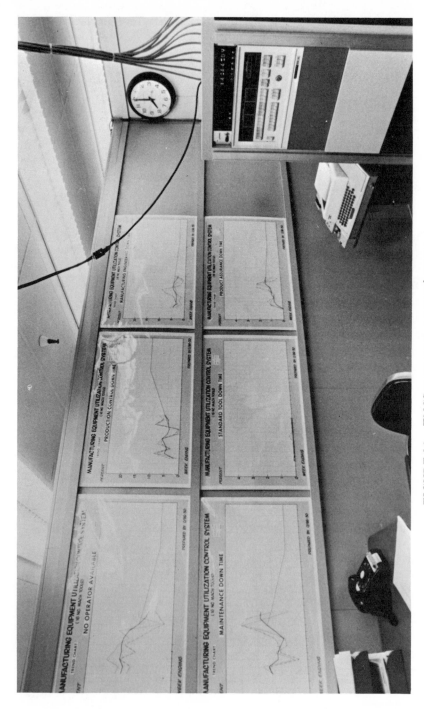

FIGURE 10. TMCS management control center.

6

Systems Engineering for Automation

Much has been written about systems engineering during the past few years. Some of this material has given the misleading impression that systems engineering is a radically new concept that has revolutionized manufacturing engineering. Of course, this is not the case. Fundamentally, systems engineering is an explicit formulation of tried and true principles of problem solving as applied to large, complex manufacturing systems.

Everyone has had the experience of expediently utilizing the most obvious solution to a problem, only to find at some later time that a more comprehensive view of the problem yielded a more effective answer. In such a circumstance, the usual conclusion is that had time been taken to analyze all the relevant factors of the problem before seizing on the most obvious solution, time, effort and, perhaps, considerable expense would have been saved in the long run. Such is the essence of systems engineering. It is a discipline that demands viewing a problem in its widest relevant context so as to identify all the

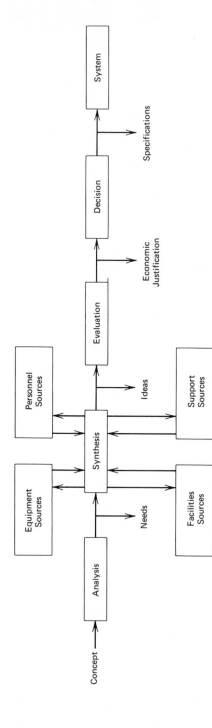

FIGURE 1. Systems engineering procedure.

factors and interrelationships involved before launching into what may prove to be a short-sighted solution.

In the integration of the various operations of a production process, systems engineering is imperative. The success of an automated production system depends upon identifying all participating elements and factors, knowing how those elements and factors are related to one another, and understanding how they can be coordinated for optimum operating results (see Figure 1). Systems engineering provides the coordination necessary to bring a complicated production process under effective, profitable control.

INTERDISCIPLINARY NATURE OF SYSTEMS ENGINEERING

A formal plan of attack is not necessary when undertaking the solution of simple problems. There are not many elements involved in such problems, the factors and interrelationships to consider and evaluate are relatively simple, therefore, there is a minimum opportunity for error. As problem complexity increases, however, problem elements and interrelationships multiply, and the chances of faulty analysis go up correspondingly. A formalized approach to solving such problems forces planners to consider possibilities that may have escaped a less rigorous treatment.

Production problems used to be relatively simple. The technologies and skills involved were restricted. But technological progress has brought increasing production sophistication, and this sophistication has resulted in specialization. Whereas specialization resulted in more powerful production techniques, it also required the collaboration of more people from more varied areas of study and application to design, build, and, oftentimes, to operate the resulting systems. Today's automated production systems include mechanical, hydraulic, pneumatic, and electronic components requiring the disciplines mechanical engineering, physics, electronics, mathematics, economics, and human engineering to design and build. Systems engineering, therefore, facilitates the management of an interdisciplinary team of experts in the creation of large, complex manufacturing systems.

Large complex production systems were first required on an extensive scale in the manufacture of sophisticated weapons systems for the U.S. military. In studying operations today from this point of view, one must raise some pertinent question: How can in-process inventory be reduced to a minimum? Can work flow be so designed that large storage areas filled with finished parts are not needed? How can the key control operations be synchronized so as to avoid build-up or "queuing up" at production centers? Inherently, this system evaluation from original order entry to final product shipment must maintain quality of product produced while operating within necessary cost objectives.

To understand the necessity of a total systems approach in manufacturing today, consider as an example the magnitude of the problem of starting from scratch to produce an automated electric motor production plant.

Without a systemized plan of attack for undertaking such a project, the prospects for its eventual successful conclusion would be dim. *Such a plan must bring to bear on the development of each part of the complex system an understanding of the objectives and needs of the system as a whole.* By dividing the overall project into subprojects, and coordinating their respective activities toward the same end, a foundation can be laid for the eventual integration of the various "pieces" into a finished system.

Furthermore, the systems concept is inherently evolutionary in nature. Thus, on-going long-range projects can be initiated with the intention of integrating specific parts of the overall project into existing operations as they mature. In this way, selective aspects of a total automation concept can be introduced into the manufacturing environment on a step-by-step basis. Such "piecemeal automation" has several advantages, including:

1. Each element of the total automation system can be fully tested under actual conditions before it is expected to perform up to full capacity as part of a totally new process.

2. Costs of introducing large-scale automation will not be an "all-at-once" proposition. They can be defrayed over a reasonable period of time, depending on when the advanced production techniques are needed, the financial state of company, and general business conditions.

3. Management and employees can get used to the new techniques and equipment at a reasonable pace, building their expertise and confidence as the systems project progresses. Over a period of time, the entire plant may be automated. Yet, no single step of the way is too big for the particular company involved to handle.

WHAT IS A "SYSTEM"?

The word, "system," is often used inaccurately. So, when considering the nature and use of the systems engineering concept in the creation of automatic manufacturing systems, it is appropriate to recognize the general character of systems.

Every system is a subsystem of some larger system and, in reverse fashion, every subsystem can be considered to be a system with its own subdivisions. Thus, the designation of specific systems and the division of those systems into subsystems for analysis, planning, and engineering is arbitrary. Of course, each subsystem should be in a single geographic location. Also, it should either totally include or totally exclude those aspects which will interface with another subsystem. The more expertly the task of subdivision is performed, the easier it will be to assign responsibility for each subsystem to independent work groups, and the less feedback will be required between these groups to achieve successful results.

APPLYING SYSTEMS ENGINEERING TO AUTOMATION PROJECTS

Applied to manufacturing automation, systems engineering is an approach for selecting and utilizing the most profitable combination of men machines, and materials to effect the optimum production performance. This approach identifies the available alternatives, formulates and uses a model of the process to compare those alternatives in relation to a quantifiable objective, and tests the chosen alternates to determine which combination will be the most effective. Systems engineering brings the rigor of the scientific method into the manufacturing environment.

The application of a formal systems engineering program to a specific manufacturing automation project can best be accomplished via a step-by-step process which blends technology and economics in equal measure (see Figure 2).

☑ Market analysis pinpoints product requirements.

☑ Project is conceived for automated manufacture.

☑ Engineering management selects project team members from permanent project administration, research and development, and product design departments.

☑ Project coordinator is appointed.

☑ Support personnel are identified from staffs of permanent departments.

☑ Arrangements are made to tap the resources of other company divisions if necessary.

FIGURE 2.   Staffing a systems project.

The following five steps are basic to any systems engineering program:

1. *Project conception.* Projects come into being because some person or group of persons recognized the need for a new system (see Figure 3). Converting this recognition into a system concept to be implemented can be a tough process. First, higher management must be convinced of the need and must be willing to

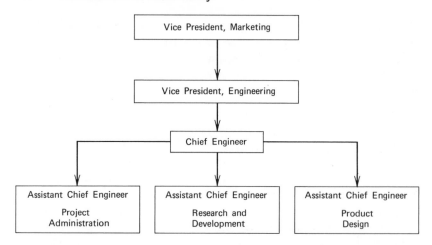

FIGURE 3.   One way to structure a systems engineering division.

make the effort necessary to produce the new system. Without solid management backing from beginning to end, failure is assured. Second, a determination should be made of the availability of funds and manpower. It will do no good to develop the concept for a grand automated production system only to discover that the resources are not available to turn the idea into reality. Third, specific objectives should be defined, and basic standards and guidelines should be established and quantified.

Quantification makes it possible to know if and when standards and objectives are met. In this stage of the project, efforts should be made to be explicit and to require the testing of all subsequent stages in relation to measurable factors. Fourth, a realistic completion date should be determined. Once the project has progressed this far, a *project or systems engineer should be assigned* to oversee and coordinate the proceedings. If the project engineer is trained in the techniques of project management and systems integration, he will be able to prevent the duplication of effort and lack of communication between participating groups that sometimes accompanies a less centralized management organization.

2. *Project planning.* This is a critical step. Here, the project must be defined in terms of system requirements such as physical system characteristics, and performance standards. If this step is to be successful, a variety of tools must be brought to bear on the problem. For example, Program Evaluation Review Techniques (PERT) and Critical Path Methods (CPM) will aid in placing key events in the development of the manufacturing system in their proper perspective. Consultation with outside specialists can bring a cross-fertilization of various disciplines and ideas to the systems project. Computer simulation

techniques may be necessary because of the complexity of the proposed system, the variations in possible operating condition, and/or the great number of possible approaches to the system's design.

Simulation requires the creation of a systems model which approximates the real world of operating conditions and will yield answers in terms of previously defined performance criteria. The purpose of the model is to test alternate system approaches in the context of anticipated conditions, without having to implement each alternative to see what the actual results would be. In any event, once system requirements and performance standards are determined — and once a basic approach is decided upon — the overall system must be divided into appropriate subsystems for subsequent individual development. Each of the subsystem alternatives must be identified, carefully studied, and evaluated. The goal is to optimize selections based upon such factors as cost, savings in operator's time, technical feasibility, and availability of the necessary components and/or materials. As selection is made from the various subsystem alternatives, system parameters are fixed accordingly.

3. *System design and development.* At this stage, each subsystem undergoes a preliminary design in which human as well as technical factors are considered. After preliminary design, specialists should be called in to provide the finishing touches, and subsystem prototypes should be produced. Prototype development will probably uncover further weak points in the design, in which case changes should be incorporated immediately.

4. *System construction.* With all subsystem development work funneling into the same terminus point, it becomes the responsibility of the project engineer to keep all activities on schedule. If one project team falls behind in developing their subsystem, the entire project will be held up, adding to the cost and confusion. By the time the construction phase of the systems project is ready for implementation, prototype development should have finalized the design of the first subsystems that will go into production. Coordinating all of the various activities involved, and resolving all of the "people problems" that usually arise, is a demanding task to say the least. Much of the success of the final system, therefore, will depend upon the expertise, decision making capability, and leadership of the project engineer or engineers.

5. *System operation.* The proof of how effective the systems engineering project was is in the nature of the results obtained. Once debugged and in operation, the new system should meet or surpass all of the performance criteria established. One of the last problem areas to be resolved is perfecting the man/machine interface. If the operators have been involved in the project from the beginning, they will have an attitude of acceptance toward the new system. Such an attitude will facilitate the start up, debugging, and early operation stages

of the project. Needless to say, all employees will resist change to a certain degree. But if the new system operators have been made an essential part of the endeavor – if their opinions have been asked for when relevant – the problems of early system operation will be technical rather than human in nature, and minimal in number and magnitude. To be able to conclude that the man/machine interface of the new system has been perfected, the following three conditions must be satisfied: first, the operators must be *able* to perform the tasks necessary to operate the system effectively and efficiently. Second, they must *know how* to perform those necessary tasks. Third, the operators must be *willing* to perform those tasks. The first condition has to do with the effectiveness of management's program of informing and training the operators. And the third condition is relevant to how well management has motivated the operators to learn and take over responsibility for the new system. The same three conditions should be satisfied in relation to maintenance, production control, production engineering, and supervisory personnel.

At the completion of each of the foregoing five steps or stages, the system can be turned over to the operating and production engineering group. Or, as is the case in some companies, the systems engineering group may assume production responsibilities. These people are the most familiar with the system, and they would be the best ones to launch the new system into regular operations.

Because of their inherent complexity and the tight integration of their various elements, automated manufacturing systems are made to order for the systems engineering approach. Automating a certain process may affect a number of operations before and after that process because of the changes in procedure or product flow brought about by automation. Systems engineering spotlights the various interactions between processes, and defines how procedures in the entire manufacturing line should be upgraded to make automation work effectively.

Furthermore, mistakes on automation investments may not be easily or economically remedied. Automatic machinery is not only expensive, but equipment designed for one purpose or process can rarely be profitably used elsewhere. Thus, it can be disastrous to undertake the development of automation via a "bits and pieces" approach. By applying the rigor and comprehensiveness of the systems engineering approach, there will be much less chance of costly blunders.

CONCLUSIONS

Essentially, automation is the integration of automatic machines. But machines cannot think whereas in a manual process an operator may spot and correct an unplanned occurrence, automatic machines will continue to function, turning out products, oblivious to any malfunction or error that might have crept into the process. Thus, automation necessitates the elimination of as many nonstandard procedures as possible — which means that all segments of the automatic process must be clearly and unambiguously specified and programmed. The only way specification and programming can be accomplished on such an exacting and comprehensive scale is via a total systems approach to system design.

In the final analysis, then, the systems approach for the creation of automated production systems is mandatory because the real paths to productivity are now in the areas of process change and improvement; because the interrelationships possible between machines and even between processes negate isolated analyses; because current market demands require a variety of products; and because only total control through the systems concept can keep production flexible enough to provide rapid changeover and short reaction time to fully and adequately serve market demands in the years ahead.

REFERENCES

"Manpower Implications of Computer Control in Manufacturing" by Arthur S. Herman, *Monthly Labor Review*, October 1970, p. 3.

"A New Look at Systems Engineering" by Robert A. Frosch, *IEEE Spectrum*, September 1969, p. 24.

The Technical Program Manager's Guide to Survival by M. Silverman, John Wiley & Sons, Inc., New York 1967.

Systems Engineering for the Process Industries by T. J. Williams, McGraw-Hill, Inc., 1961.

"Erase Old Organization Chart to Make Systems Concept Work" by G. J. McManus, *Iron Age*, October 22, 1964.

"Systems Can Too Be Practical" by Allan Harvey, *Business Horizons*, Summer 1964.

"How Does Systems Engineering Fit into the Process Control Picture?" by J. Allan Wickett, Jr., *Chemical Processing*, September 9, 1963.

Industry Case Examples

CASE 6A

Automation Solutions in Electronic Production

The Situation. A large manufacturer of electronic components is faced with the problem of mass producing large quantities of small coils for magnetic-latching reed switches. The coils are wound on plastic inserts in a steel plate on an eight by eight matrix. The inserts extend from both plate surfaces and provide 64 winding cores on each side. Each core supports a primary and a secondary winding for a total of 256 windings per plate assembly.

Automatic winding machines are being used in production runs, but they do have severe limitations. These machines are largely electromechanically controlled with a maze of gears, levers, and approximately 60 to 70 electrical switches that must be adjusted or set for specific coil winding patterns. The switch plate matrices are wound in eight different patterns and skilled operators are required to make the necessary adjustments whenever a pattern program is changed.

Basically, the electro-mechanical winding machine consists of a table that moves in X or Y ordinates under four coil winding spindles. Two switch plates are mounted on the table for each coil winding run; they are mounted in such a fashion that two of the spindles wind coils on one plate while the other two wind coils on the second plate.

Drive motors, controlled by cam-operated switches, sequentially position the table under the spindles to wind each coil. While each spindle is wrapping the wire around the circumference of a plastic coil core, a third motor is driving a cam which lifts the table so the coil will be wound in a helix from the top of the plastic core to the plate surface. Mechanical stops and switches control the three axis of table movement.

Since the underside of the table is blocked by mechanisms, the coils can be wound only from the top side of the machine. The pattern on one side of a switch plate must have a matching mirror image on the opposite side, therefore the geometry of the winding machine is such that two plates being wound simultaneously are wound in mirror image patterns to one another. When the opposite sides of the plates are to be wound, they are turned over and their positions on the table switched so that each row gets a mirror image pattern of its own previous winding.

132

Mechanical aspects of the electro-mechanical winding machines permit some coils to be wound with overlays or voids which result in rejection for the affected switch plate. Although the switch plates can be salvaged by rewinding the faulty coils, it is a costly and time consuming process.

Added to the high rejection rate, more time is lost in the necessity of performing hand finishing on the wound switch plates. Limitations in the electro-mechanical programming make certain automatic finishing procedures impossible. Some wires have to be manually trimmed, others have to be hand wrapped around stabilizing pins, certain connections must be soldered in by the human operator.

Seeking to make the electro-mechanical machines more truly automatic, hoping to improve the quality of the finished switch plates, and desiring to reduce the required level of operator skill, the manufacturer sought improved production techniques and equipment. His search led to electronic automation.

Automating the Situation. Using a General Automation SPC-12 automation computer, the manufacturer replaced the electro-mechanical controls of the coil winding machine with a program stored in the computer's core memory. The program contains all eight coil winding patterns and, because of its increased flexibility, consists of far more detailed instructions than possible with the electro-mechanical system. Manual hand finishing is practically eliminated from the production procedure.

Computer initiated drive improvements also have reduced rejections to almost zero.

The System. Communication with the automation computer is through a teletype keyboard or a paper tape reader. When desired, data readout is displayed on the teletype.

Program output from the computer is directly to the coil winding machine's various drivers; program input is from the machine's position indicators. Output commands are compared to input results in a closed loop cycle.

Three switches replace the former maze of nearly 70 adjustment and control switches in the electro-mechanical winding machine. One switch for on/off and one each for the X and Y table position over-rides. No cams, levers, stops, or other mechanical adjustments are necessary with the computer automated system.

Because of the high response rate of the computer it is possible to replace the cumbersome cam-operated vertical table drive with a more positive direct-drive pulse motor. The computer can easily count the rapid spindle encoder blips and pulses the vertical drive 20,000 times a minute. Since the spindles are winding at 1,000 rpm, this provides 20 pulses per wrap. Each pulse moves the table 0.001 and, since the wire is 0.0020 in diameter, the movement wraps the wire in a tight helix. Voids and overlays are eliminated.

Operation. Programs are loaded into the automation computer's core memory through the paper tape reader. Prior to a production run, the shop lead-man decided which pattern is to be run and selects it via a teletype keyboard coded entry.

When the operator wants to start the run, he pushes the start button at the machine. The start signal is sensed by the automation computer which then takes over control of the machine. At the end of the run the computer shuts down the machine.

The switch plates are manually loaded and unloaded by the operator which, other than pushing the start button, is about all there is for him to do in the procedure. He may also initially move the table close to the starting position before turning the machine on so the computer will not have to slew the table into the starting position. Manual movement of the table is accomplished by the X and Y control switches on the machine.

The great flexibility and program capacity of the computer control system will permit simultaneous control of more than one machine. In fact, the manufacturer has another system in development that will control sixteen machines; all machines can be running the same pattern; or any combination of machines and patterns is possible.

As far as the operators are concerned, the expanded system will be as simple as before where only one machine is controlled by the computer. Each operator will be able to handle two machines at a time, loading or unloading one, while the other is running. The shop lead-man's job will be slightly more complicated by the fact his inputs to the computer must now identify which patterns will be run on which machines instead of merely selecting a single pattern for the solitary coil winder.

The computer will monitor all the on/off switches and channel the selected pattern to each machine as it senses it has been turned on.

Advantages. Although the winding machine does not run any faster because of computer automation, the production rate is significantly increased as hand work is practically eliminated and the rejection rate is reduced to nearly zero. Both of these factors also significantly reduce the per piece costs of the finished switch plates.

Computer programming takes all of the tedious manipulation of the adjustments out of the hands of the operator and this, along with the elimination of hand finishing, permits operators to perform more effectively. In fact, in a test, one inexperienced operator on the computer controlled machine out-produced two seasoned operators on a pair of electro-mechanically controlled coil winders.

CREDITS: General Automation, Inc.

CASE 6B

Automating Vertical Traffic

Haughton 1092-IC applied systems technology in order to create an entirely new approach to the elevatoring of high rise buildings. As advanced in action as it is in looks, 1092-IC is the product of eight years of electronic development and three full years of operational success.

Integrated circuits provide infinite capacity and complete flexibility with millisecond speed for enough electronic logic to supervise and weigh every factor affecting elevator service (see Figures 4 and 5).

The ability to treat each passenger requirement as a separate obligation distinguished 1092-IC. Every factor influencing passenger service is sensed and responsed to instantaneously. Service is optimized by keeping track of what each car is doing, comparing it with the availability of every other car, then based on that assessment allot or re-allot calls in milliseconds to achieve the optimum service.

The result of this technology: the shortest Destination Time (the total time after pressing the hall call button to arriving at the destination floor) ever achieved in group service. This seems not too demanding until projected in an example: A fifteen-story building has four elevators (see Figure 6). There are eight corridor calls to be answered. Considering all influencing factors there are over 27,000 possible combinations of calls-to-cars assignments — only one of which would afford the shortest wait and fastest travel: Optimum Destination Time. A roomful of conventional relays used on other control systems would theoretically reach the right decision in about 90 seconds. 1092-IC does it in ten milliseconds. And 1092-IC re-examines the service every time new information is available.

The entire elevator plant becomes more efficient than ever before, because now any car can be deployed anywhere to suit passenger traffic. And 1092-IC accomplishes this in less than the former space, with greater reliability.

CREDITS: Reliance Electric Company, Haughton Elevators.

FIGURE 4. Automatic elevator controls showing IC digital logic printed circuits for dispatching elevators.

FIGURE 5.   Car alloter control panel.

FIGURE 6. Modern elevator automated with electronic logic optomizing control system for 15-story building.

Management Practices and Policies for Automation

7

Establishing a Management Philosophy

Equipment and techniques comprise the technology of automation. But when should this technology be applied in specific instances? To what extent? And how can automation technology be used to foster the overall growth of a company toward long-range goals? To answer these questions in any particular business context requires knowledge, analysis, decision making, and planning. These are the functions of management, and management is responsible.

The functions of management are so basic that they are *often* taken for granted. Consumers, for example, are so used to an unlimited choice of products that fulfill so many basic needs and desires, and do so many amazing things, that they do not stop to think how it all comes about. Without thinking, they assume that ideas are automatically "plugged into" technology to produce everything from safety pins to hundred-story office buildings and 747 jets. They do not comprehend that every tool of production — and every product produced by it — had to be conceived, developed, and systematically produced, marketed, and upgraded at a *profit*. The thought, skill, effort, and planning needed to bring this all about do not occur by chance. Competent management is the most basic and most vital ingredient in any successful enterprise. Automation technology makes the manager's job tougher — and his prospects for success greater!

THE NEW MANAGERS

One often hears proclaimed, these days, that the managers of today and tomorrow must be basically different than their predecessors. Some savants claim that modern technology has made the traditional concept of management obsolete — that managers of the future will be more adjuncts or overseers of the electronic computers that will take over the decision-making and planning functions which used to be reserved for humans. Others see the task of management getting too tough and complex for individuals, in which case decision-making committees would replace individual department heads and executives.

Neither of these views is really valid. Unless and until robots are created that are essentially indistinguishable from their human counterparts, machines will always need someone to conceive, build, maintain, and integrate them into a productive system. And, most important, they will always have to be programmed and utilized in accordance with human values! What *is* happening, however, is that the increasing capability of machines — especially computers — is transforming much of what used to require human decision-making into mechanized calculation and/or manipulation. This trend of preprogramming more and more sophisticated "software" functions into the hardware will continue. But it will not render managers obsolete. To the contrary, it will free the manager's mind for the larger perspectives and more complex decisions that automated production requires.

Committees as well can neither replace nor supplant individual decision-making and responsibility. If a business organization is not to become bogged down in the bureaucratic maze of "consensus-making," it must, in the long run, rely on the initiative, expertise, and responsibility of competent individuals. Committees have their place, but *not* as the arbiter of new ideas or key decisions.

What qualities must the new managers possess if they are to be successful in the utilization of automation to advance productivity? Tomorrow's managers must:

1. *Study and understand the fundamental principles of automation technology*. It is fashionable these days to depict managers as corporate psychologists dealing only with "people problems," while "mere" technical considerations are

delegated to "underlings." Don't you believe it! Although it is true that a competent manager must know how to deal with people in an effective manner, he must also have enough of a grasp of technical matters so as to be able to evaluate key problems and proposals. A manager can delegate authority, but not responsibility. Ultimately, *the manager is responsible for the success or failure* of an automation project. Without a fundamental understanding of what automation is all about, the manager will be at the mercy of those who work for him. While it is true that one man can't be an expert in all of the specialties necessary to an automation system, a manager can be competent in the principles of how those specialties must be organized so as to produce the most successful result.

2. *Develop an objective attitude toward change.* Change is neither intrinsically good or bad. The object of applying automation technology is not to "keep up to date." Neither is there any virtue in maintaining traditional ways of doing things. Change is progress only if it aids the realization of specific goals. In terms of automation, this means that new techniques and equipment should be introduced just as fast and to the extent that is necessary or desirable to move toward manufacturing and/or company objectives. Managers should develop the attitude of being open to progressive change — change which will advance the productivity and also the profitability of the company.

3. *Welcome automatic machines as advanced tools to be applied to complex tasks.* Automatic machines are neither demons to be feared nor "partners" to be "trusted." They are complicated, highly-capable tools — but tools none-the-less. Used properly, they can mean the difference between success and failure or between rapid and slow growth. For example, a manager who. in principle, rejects the use of a computer in making decisions will not only be less effective in what he does, but he will not be able to put his attention and efforts where they belong: toward increasingly more creative tasks. The result will be that his company will fall behind those that have understood and utilized the computer's potential for increasing management productivity. All other factors being equal, a good tool is not as effective as a better tool. Managers who, for some reason, have developed an aversion to sophisticated, automatic equipment will, in time, discover that they have undermined the future of their enterprises.

4. *Know your products and markets.* This is sound advice for any manager in any business situation. It is even more imperative when such knowledge will serve as the basis for multi-million dollar investment programs that will introduce automatic equipment, multiply product output, and/or reorganize manufacturing facilities. The more a manager knows about his company's products, markets, and operations, the better his evaluations and decisions will be.

5. *Learn and teach the objective values and economics that constitute the basis of a free market.* Immersion in the day-to-day activities of running a

business often results in a manager losing sight of the larger social/economic context of which his enterprise is a part. Individual rights, private ownership of the means of production, service and competition — these are the pillars which support any business pursuit of profit. Without a thoroughgoing knowledge of the principles of and the alternatives to free enterprise, management can only apologize for business activities — activities that have produced more products, employed more people, and created higher living standards than anywhere else in the world at any time in history! These benefits cannot be maintained for long without the ideological underpinnings that made them possible in the first place. No manager can consider himself competent to deal with the new technology without identifying and advocating the social/economic principles that led to its development.

CREATING A TOTAL MANAGEMENT PHILOSOPHY

How does management go about organizing its functions and activities so as to maximize its effectiveness in an age of automation? Recognizing that social/economic forces are intensifying the trend toward automated production, how can management gear itself to the task of intelligently using automation technology to spur its company's growth?

The answers to these questions lie in the nature of the automated production system itself. Such systems come about through the automation and integration of various functions. The result is that what used to perform in segmented parts now operates as an optimized whole. This change in operation on the production floor requires a corresponding change in perspective in the executive suite. The systems approach to manufacturing must be complemented by a "total-philosophy" approach to management.

What is a total management philosophy? It is the explicit definition and bringing together of what were, traditionally, separate management functions for the purpose of being able to perform on the managerial level with the same quickness of response and effectiveness that automation has made possible on the production level. All the old questions need to be re-examined — and some new ones asked — in the light of the increasing need for automated production systems. Problems to which management formerly reacted in expedient fashion must now be anticipated and planned for in advance. Opportunities that seemed to appear by chance must be systematically searched out and implemented. Only a total, integrated philosophy will give management the overview required to perform in this fashion.

The place to begin is with a thorough study of the manufacturing process in its entirety. This new attack on the problem contrasts sharply with the departmental or piecemeal approach which has been common in the past. In developing and implementing this approach, it is imperative to recognize that an automated production system consists of four basic functions (processing, handling, control, and data processing) that are inextricably woven into the final whole. The evolution of automation technology in any manufacturing context is toward increasing greater integration of these functions.

Management development programs that focus on how the new technology is changing manufacturing and decision-making should be used as the major tool for educating company managers and supervisors. Such a program will assure that automation will exert its proper impact upon the thoughts and actions of those who will be guiding the future growth of the enterprise.

Implementation of a total management philosophy will necessarily require some reorganization to avoid traditional internal departmental barriers. One of the most important facets of reorganization will be the creation and utilization of a systems engineering approach (as discussed in the previous chapter) so as to maximize the effectiveness of every change toward more automation. Management must learn to think in terms of systems if systems engineering for automation is to produce the results desired. Once automation technology is systematically applied to raise the productivity of manufacturing operations, all management decisions must be based upon the interrelatedness of production functions.

MANAGING PRODUCTIVITY EFFORTS

American industry for many decades maintained its competitive position in the world markets through its ability to continuously improve productivity in its manufacturing processes. A combination of aggressive capital investments coupled with improvement in worker skills made it possible to maintain an average gain approximating 3 percent annually. The records indicate, however, that our productivity gain has not done so well in recent years. In fact, there are indications we have had no gain at all.

In the meantime, our competition such as Germany, Japan, and others have fully recovered from the devastation of war and with new plants are taking full advantage of new technology and the total systems approach mentioned earlier. This, along with the employee enthusiasm for work, has been yielding productivity gains far in excess of those in American industry.

In addition, increasing consumer demands for more various and complex products, increasing government regulation of private enterprise, and consistently rising labor and material costs places a higher premium on a company's ability to increase productivity. Increasing use of the latest automation technology is a must if one is to meet these pressures. It must be a constant improvement of productivity to maintain profitability. Productivity is what automation is all about, and automation requires creativity in development.

Managing automation in the years ahead means that the introduction of necessary and/or desirable change must *not* be left to expedient, random processes. Management must be constantly encouraging, developing, testing, and employing new production ideas, techniques, and equipment.

Formulating and implementing a management automation policy requires a grasp of what this innovation entails. Basically, innovation here is the conversion of new ideas and approaches into some new, useful production technique, equipment, or product. The dual objectives of an effective automation policy, therefore, are (1) to create a work environment in which creativity can flourish and (2) to establish procedures for evaluating and implementing pertinent ideas and suggestions.

In regard to fostering this creativity, remember that there is no substitute for creative personnel. You can't simply alter an essentially noncreative person's working environment and, thereby, transform him into an innovative thinker. The key is to recruit for creativity when staffing those departments of the business where creativity is important. And once those creative individuals are on the job, give them the freedom and backing necessary to encourage the generation of new perspectives and techniques for accomplishing existing tasks.

Where evaluation of new ideas and techniques is concerned, be sure that the nature and scope of a specific suggestion and the total context of its implementation is fully understood before judgment is passed. Often, management decision-making on this score is uninformed and/or arbitrary. Not only will such a procedure gloss over potentially profitable ideas, but it will discourage further creative efforts as well.

Three conditions are necessary in a company to assure successful creativity with regard to automation:

1. *Recognized opportunity.* This opportunity may not embody a "need" in the sense that a problem exists that must be solved. Many very profitable improvements can be made in situations and/or operations that are already functioning adequately. The criterion for implementing a new idea or introducing new equipment should not be whether or not the action eliminates a problem, but, rather, whether or not it results in an improvement sufficient to justify the effort and capital required to bring it about.

2. *Competent personnel.* The most opportune situation coupled with the best idea still is not sufficient to result in profitable automation. The idea must be reduced to a plan and/or hardware, and this requires the efforts and abilities of skilled craftsmen and highly trained engineers. A good idea incompetently implemented can turn out to be worse than an idea with less potential that is

competently executed.

3. *Financial support.* Once the decision has been made to introduce a new technique or system, don't "pinch pennies" in implementing the project. Be prepared to miss or exceed the original estimate of return on investment. The error should be relatively small so that the gain is worth the investment. The purpose of automation is to provide a lasting improvement in production operations. Don't compromise the success of the innovation by cutting corners that will have detrimental consequences later. If adequate financial support is not available to do the job right, shelve the program and re-evaluate it at some later date.

Success with automation cannot be "guaranteed" any more than creativity can be taught. But a forward-looking management can maximize its opportunities for profitable automation by a systematic and continuous reappraisal of the way the business is operating and of the direction it should be taking. Good management does not only involve creative adaptations to new conditions; it means the constant creations of new conditions as well. Innovative automation is purposeful, organized risk-taking change introduced for the purpose of increasing profitability. As such, it is as important in policies, goals, organization, marketing, and communications as it is in the technological areas of product and process.

SETTING GOALS

What other steps can be taken by a progressive management seeking to promote automation? The following possibilities should be considered:

1. Set goals that stretch the imagination and tax the efforts and abilities of management, supervisory, and staff personnel. Though some part of every work activity will inevitabley be mundane, the aims of each department and the company as a whole should be such as to require continually good performance to achieve.

2. Institute and adequately maintain vigorous research programs. Today's research is the fountainhead of tomorrow's technology. Staff research programs with the best people available, and reward them according to their performances.

3. Dramatize a "fresh ideas" company image to people inside and outside the company via employee communications media and advertising. The best talent — both young and old — thrives on challenge. By maintaining the image of a company flourishing on the creative achievements of its employees, management will set in motion a selection mechanism that will attract and keep exceptional performers.

4. Practice management innovation by constantly evaluating new management aids and techniques, and utilizing them where desirable. No longer can management sit back and direct changes in production operations. Production is becoming increasingly complicated and sophisticated. Management must employ new technology to its own efforts if it is to stay ahead of the game.

5. Create special executive task forces to examine key areas of challenge, trouble or opportunity and to recommend step-by-step programs for attacking them. Don't try to solve major problems or take advantage of prime opportunities with a "business as usual" approach. Special circumstances deserve special efforts. Such an approach will stimulate interest among participants and other employees, as well as offer the best hope for effective handling of the issue in question.

6. Reward and promote employees according to performance. The only way to assure future company growth is to continually attract and keep bright, young talent. The steady influx of fresh thinking into the management ranks will not allow the creative spirit to die in a company.

7. Establish a permanent idea group to continually explore the frontiers that lie just beyond present company policies and activities. Such a group would function as an "early-warning net" in regard to possible *changes in the marketplace* and potentially useful methods and devices emerging from basic research. In addition, this group would formulate new programs to progressively introduce into company operations as conditions warrant.

Though the widest possible latitude must be given to the generation, evaluation, and introduction of new ideas in a company, management must learn to distinguish between new ideas and speculation. All suggestions for improving and/or renovating production operations must be gaged against the discipline of "probable profit potential." Some of the major reasons why attempts to innovate fail to create profitable end results can be listed as:

1. Over-estimation of the technology available to implement particular automation projects. Such miscalculations result in compromises and/or program turnarounds that have severe economic consequences.

2. Preoccupation with sophisticated solutions. The aim of any automation project should be the economical improvement of production operations. If complicated methods and/or systems are necessary to accomplish this under particular circumstances, so be it. But there is no advantage in sophisitication and complexity, per se. Many times the best solutions are the simplest ones.

3. Unrestricted growth of technical support groups into "islands of speciality." A manufacturing company is not in the business of basic research for its own sake. Special, technical disciplines are to be used to continually improve productivity. Care should be taken to direct all efforts toward that goal, and not allow separate, scientific power bases to develop within the company.

4. Inability to distinguish between basic scientific information and true technological development. A technological development is useful toward some profitable end — if it is economically feasible. There is much scientific information that does not fit that category.

PREPARATION FOR IMPENDING CHANGE

How should an enlightened management go about cushioning the effects of introducing automated systems into production operations? What steps can be taken to assure that employees will aid, not fight, the planning, installation, and operation of a new system? The following points apply:

1. Be sure management — from the supervisory level all the way up to the top executives — is fully committed to the project and will participate in planning its introduction. Without a unified front, management will succeed only in generating more fears about the effects automation will have upon company personnel.

2. Fully inform all employees about the change on a "need-to-know" basis. Those workers most affected by the new system should receive the most information and counseling. What kind of information should be communicated? First, explain the reason for the change (customer requirements, general business conditions, etc.) and why the chosen course is the best under the existing circumstances. Second, describe the step-by-step details of the plan, pinpointing which employees will be affected most, how, and to what extent? Third, give a comprehensive presentation of the benefits that will be realized. Fourth, explain the steps management is taking to minimize the impact and/or adverse effects of the change on employees. Fifth, outline where and when unavoidable work dislocations, delays, or problems will take place, and enlist worker support to help make the transition easier. Sixth, present periodic progress reports on the problems encountered and the solutions implemented. Seventh, don't forget to express your gratitude for the efforts employees have exerted in helping to make the change and achieve expected results.

3. Be sure your communications efforts are effective. Good communication should be timely, thorough, specific, candid, participatory, and repetitive. Most communications take the forms of informal oral, formal oral, visual/oral, or written. Use them all.

4. Be ready for emergencies. Managements that *generally* do a good job of preparing employees for major technological changes will have a solid foundation from which to handle the unexpected. When the unexpected occurs, accelerate normal procedures and enlist top management in a more direct, immediate role. Always keep in mind that good performing. loyal employees are

worth their weight in gold. Any policy that minimizes their value in a crisis will have severe, long-term effects when the crisis is over.

CONCLUSIONS

What will be gained by gearing management policies and decision-making to the requirements of automation technology? First, and foremost, the new technology will be accepted and used as a prerequisite for progress. Continued R&D and innovation management will lead to a stable, long-range capital equipment program that will assure company competitiveness well into the future. Second, effective communications will have been established with employees. They will be secure in the knowledge that they will be fully informed on matters that affect them, and will be given every opportunity to retain or advance their jobs with the company in the event of a major change in production methods. Worker morale, loyalty, and productivity will improve. Third, existing employees will provide a pool from which to make long-range manpower projections to match automation planning. Retraining programs will assure that management will not be totally at the mercy of technical personnel markets whenever it decides to make a major change in the technology of its plant.

Fourth, a record and image of progressive success will have been established that will spill over into the community. Take advantage of this by arranging a continuing public information program. Promote local community and industry understanding of your automation planning. Communicate and promote your "people planning." Explain your plans for countering outside threats to your business and your employees' jobs. And demonstrate how these policies are in the best interests of everybody concerned – management, employees, and the people of the surrounding community.

REFERENCES

Automation and Management by James R. Bright, Division of Research, Harvard Business School, Boston, 1958.

Managerial Innovations of John Diebold by M. S-C. Henderson, LeBaron Foundation, Washington, D.C., 1965.

The Shape of Automation for Men and Management by Herbert A. Simon, Harper & Row, Publishers Inc., New York, 1965.

Automation Management – The Social Perspective by Ellis L. Scott and Roger W. Bolz, published by the Center for the Study of Automation and Society, Athens, Georgia, 1970.

Transition to Automation by Otis Lipstreau and Kenneth A. Reed, University of Colorado Press, Boulder, Colorado, 1964.

Computer Technology – Concepts for Management, Papers presented at a symposium conducted by Industrial Relations Counselors, Inc., Greenwich, Connecticut, May 28, 1964.

Industry Case Examples

CASE 7A

Automated for Safety

Preventing costly oil and gas spills in the Gulf of Mexico is the job of a silent watchmen being used by Texaco. An automated shutdown system, fabricated by Robertshaw Controls Company (Brown & Root, Inc., engineering contractor), is installed on a multiple-well gas production platform located four miles offshore in frequently storm-tossed Atchafalaya Bay, near Lousiiana's oil-rich coast (see Figure 1).

The bay is wide and shallow, permitting winds to churn the water into a platform-wrecking frenzy "almost instantaneously," as a Texaco engineer put it. Guarding the entire complex of fifteen gas distillate wells which feed the platform, plus a pipeline bringing in gas from a second platform, is a system of innocuous-looking green boxes and pneumatic valves which will prevent any damaging spill from happening. This system can quickly shut in any number of wells, the entire gathering platform, or any of the equipment on it, in case of emergency.

For gas distillate wells, this sort of protection is especially vital. Distillate is an extremely light and volatile oil, usually found in connection with natural gas. The Texaco platform produces 75 million cubic feet of gas per day, and hundreds of barrels of distillate, from wells under about 3,500 pounds of pressure.

But the shut-in hi-lo pressure sensors in the system make a damaging break next to impossible. Even though the flowlines can stand full wellhead pressure, if the pressure or flow rate on any well should exceed the normal operating ranges set into the control unit, it is immediately shut in.

If the gathering line which runs from any given well to the platform should break (in a storm, for example) the controls at both ends of the line would sense the low pressure condition and instantly activate valves to seal off both the well and the platform from the broken pipe. Spills are thus kept to an absolute minimum.

The same automated cut-off protection is provided for the 16-inch pipeline running from the platform to shore. Additional sensors safeguard the production equipment on the platform itself.

If a fire breaks out, there is a master shutdown system which is activated automatically. It shuts in the wells, pipelines and platform equipment

152

simultaneously, isolating everything in the production network.

The system also monitors pressure and temperature in the steam system used to prevent the gas wells from freezing as the gas expands at the surface. The gas production rate is automatically adjusted to a suitable level in the event the steam pressure or temperature drops below the normal operating range.

At any time the system is activated, a red flag drops in an indicator window on the master control panel (in Figure 2), showing which well or line has been shut in and why. Also, a siren is sounded which alerts operating personnel.

The system has come to be called "The Screamer" because, as Texaco offshore men say, "It starts hollerin' when it hurts."

CREDITS: Texaco, Inc., and Robertshaw Controls Company.

FIGURE 1. Multiple-well gas production platform near Morgan City, Louisiana, is protected from oil and gas spills by Robertshaw's "Screamer" system.

FIGURE 2. Well and platform safety shutdown and separator selector panel developed by Robertshaw for Texaco platform in Atchafalaya Bay.

8

The Finance
Policy

Guarantees that investments in automated facilities will be financially sound are hard to come by. However, some measures of a sound investment, the methods of gathering complete factual data, and some of the pitfalls to be avoided are important to recognize. Simple forms for summarizing and presenting pertinent information are desirable in order that some rational means of making comparisons is possible over a period of time.

Financial analysts consider return on total assets employed in a business organized for profit to be the best of the measures of the success of that business and of the management of the enterprise. Return on total assets is the percentage relationship of the profit earned to the total assets employed, the assets being made up generally of cash, receivables, inventories, plant, property, and equipment.

Assuming acceptance of the soundness of return on investments as a measure of success of a total enterprise, it is logical to apply that measure to that segment of the business dealing with manufacturing. The ultimate objective, therefore, of any investment in automated manufacturing facilities being considered by a company organized for profit should be to obtain a satisfactory return on the investment involved. This is true whether that investment is for a very simple machine or a complex, highly automated manufacturing line. In general, the objective for return on investment should be such as to result in an overall acceptable return on all assets being employed in the business.

These broad objectives and measure of results are likely to be readily accepted. Problems, though, immediately manifest themselves when all of the facets of any but the simplest specific potential investment are to be considered.

The capital expenditure analyst is faced with a wide array of such factors as: forecasts such as production volume, life expectancy of the product line, life expectancy of specific parts of the line, and life expectancy of the equipment under consideration; proposals covering various approaches. equipment, and processes; quotations for equipment covering all or part of the total required; estimates and "guesstimates" on costs and savings all the way from A to Z; inflation or deflation; and tax considerations.

Even if it could be assumed that the analyst could achieve perfect answers to all of the elements just mentioned, and thus arrive at a mathematically and factually correct return on the proposed investment, the question still could remain as to whether the investment would result in an over-all increase or decrease in the corporate return on investment. This is because few, if any, can know for sure the proportionate contribution to the overall return on investment that is currently being earned by any one piece or group of manufacturing facilities. The earnings "bogey" for the proposed equipment must therefore be an estimate, *not* a mathematically assured value.

AIDS TO EXPERIENCED JUDGMENT

Little real progress has been made in this field over the years because of the complexity of the problems involved in arriving at a scientific answer indicating return on investment achievable for a proposed facility. Seat-of-the-pants judgment, carefully exercised by experienced capable businessmen, has resulted in wise investments that earned good returns and consequent success and growth of their companies. Seat-of-the-pants judgment carelessly applied by inexperienced men has resulted in many business failures. Purchase of equipment for a perfectly sound application from a vendor lacking the innovative capabilities required for automation can result in equally disastrous failure.

Over the long haul, the manner in which capital dollars are invested will determine a manufacturing company's competitive position in the market, at least as much as any other facet in the operation of the business. Because of this fact, final decision on whether and when large capital investments are to be made must be the responsibility of the top executive or operations committee of the enterprise. These men cannot do the analysis of all investment proposals themselves, therefore they must of necessity depend upon the accuracy and completeness of data prepared for them by subordinates down the line. Thus, the abilities of the manufacturing engineers are of prime importance.

Training of these analysts in the several divisions of a corporation in the consistent application of an analysis procedure is of prime importance. Training is also of prime importance to the executives who must make the final decisions for approval or rejection of capital expenditure proposals. They need training in the uniform understanding of the value to the corporation and to themselves as individuals of use of analysis procedure as an aid to the application of sound business judgment in evaluating investment proposals.

In a multidivision corporation, a corporate staff group such as a manufacturing engineering department should have extensive manufacturing knowledge in the areas in which the various divisions are involved, and of their facilities and needs. Such a staff group is then in the ideal position to train the division analysts, to review their recommendations, *and to audit the performance results* obtained after installation and use of new facilities. In a smaller company, the manufacturing manager or manufacturing engineer should train the analyst or prepare the analysis of manufacturing facility investment proposals himself.

As of today there is no known formula that will crank out guaranteed answers to sound, profitable manufacturing automation investment proposals. At best, the available formulas and procedures can be considered as more or less formalized standard techniques to *aid experienced business judgment* in evaluating: (1) the merits of a proposed investment, (2) the relative merits of various proposals to accomplish a given operation, and (3) the relative merits of various proposals competing for allocation of available funds. Good, sound, experienced business judgment still holds the key to the making of profitable facility investments, just as it was many years ago before the concept of return on investment as a measure of performance had been developed.

Thus, there is a fundamental need for an understanding of the measure of worth of an automation investment, the broad financial objectives being sought, the complexities facing the analyst of the various proposals, the magnitude of the importance of making sound capital investments, and the importance of sound experienced business judgment in the making of correct decisions in this area. To provide the basis for good judgment requires development of costs and savings allied with a proposed facility investment, and the conversion of this data into a recommendation for favorable action based upon return on investment and other useful indexes of worth of the proposal.

USE STANDARDIZED PROCEDURES

Formalized, standard techniques should be used in the analysis of all capital expenditure proposals covering individual items of equipment or complete programs involving automation. For practical reasons it is suggested the procedure should be used where: (1) the total capital expenditure involved is $5000 or over and (2) the proposal is based upon cost reduction. An example is shown in Figure 1.

Such a procedure should not be used where a proposal is based solely upon intangible considerations. This type of proposal should be kept to a minimum.

When a program involves facilities for a complete new product line where no facilities currently are available, the procedure should be used for evaluating the relative merits of various proposals to accomplish each given function. Evaluations should be made by comparing the savings accomplished by each more costly proposal to the net additional capital investment required by that proposal over the least costly proposal. For the program as a whole, the return on investment should be developed by preparing annual operating statements normally covering a ten-year projection of operations.

Each automated facility investment proposal should be considered in the light of existing conditions and be evaluated on the basis of the net capital investment involved, the net capital investment being the increase in book value of assets resulting from the proposal.

AN EXAMPLE SHOWING FORMS AND CALCULATIONS

Drilling, tapping, and counterboring of the frame mounting holes in the main poles and interpoles of integral horsepower direct current industrial electric motors was currently being performed in a three spindle gang drill using guide ways and individual cap type drill jigs for locating the holes. The operation was quite efficient for a manually loaded, manually indexed, and manually unloaded cycle.

As the equipment approached the end of its service life automated means were considered in the light of existing lot quantities and forecasted total quantities. The investigation developed a special machine with improved clamping and locating means and with combination automatic and operator-controlled index and machining cycles. Considerably improved setup time requirements permit reducing inventory materially without adversely affecting machining costs. Manual loading and unloading during the machining cycle is retained.

In checking the degree of automation it was found that zero return on investment would result from replacing the worn out machine with a modern machine of the same type, since the existing setup and operating time standards would not be significantly improved. The only savings with new equipment would be in reduced maintenance and down-time. These savings would be more than offset by the added depreciation charges.

Addition of automatic loading and unloading has not been studied because the preceding operation has not yet been re-studied. Judgment indicates that automation of that operation will not likely result in economical elimination of manual unloading of that operation and loading of the drilling operation.

Sample forms illustrate data developed to evaluate a proposal for the drilling, tapping, and counterboring operations. Work Sheet I (Figure 1) summarizes the proposal data and results expected. Work Sheet II (Figure 2) records the savings and cost data and the calculation of the indexes of worth. Work Sheet III (Figure 3) records the Investor's Rate Method data and selection of the interest rate.

The net cash returns are multiplied by factors from a standard-interest factor table to get the present value of all cash returns during the service life of the proposed equipment. The interest rate at which the total of the present values of cash returns for the service life equals the net cash expenditure is the interest rate for the Investor's Rate Method.

The return on investment of 42.8 percent before tax is considered very good, particularly when an inventory reduction of almost half of the cost of the equipment can be accomplished, and when the machining method being replaced was a fairly modern, efficient method.

PRIME MEASURE OF WORTH

Return on net capital investment, before federal income tax, should be the prime measure of value to be considered by the operations executives when making decisions on capital expenditure proposals. This return is the percentage relationship between the average net operating savings before taxes, but after depreciation charges, and the average book value of the net capital investment, both averages being taken over the service life of the proposed asset.

Net operating savings are gross operating savings less the difference between the depreciation charges being assessed against the existing facilities, and the depreciation charges that would be assessed against the new facilities. Gross operating savings are the difference between all savings and costs incurred as a result of the proposed investment, when compared to the costs of manufacture with the existing facilities.

Depreciation charges are generally determined by a controller's department, based upon many factors often beyond the knowledge of the capital expenditure analyst. However, it is important that the controller's department have knowledge of the expected service life of the facilities being proposed. Only with this knowledge can the controller's department select the most appropriate depreciation schedule.

Capital Expenditure Analysis   Work Sheet I
Proposal Data and Results Expected

---

1. Application Description

    Drill & Tap Main & Interpoles

Analysis No.
Date: 10-22-59
By: JIK-KHM
Location: Ivanhoe Div.

---

2. Present Equipment Data:
   a) Description:

   3-Spindle gang drill

   b) Present Book Value $ -o-
      (Per Acctg. Books)
   c) Present Market Value $ 200
   d) Capital Gain (c-b) $_____
   e) "   Loss (c-b) $_____
   f) Current Rebuild or
      Retool Cost $_____
   g) Age of equipment   19 Yrs.

3. Proposed Equipment Data:
   a) Description:

   Automatic Drill & Tap Machine

   b) Acquisition Cost (Capital) $57,507
   c) "   "   (Expense) $ 2,000
   d) Install.& Trans. Cost $ 2,825
   e) Removal Cost-Pres. Equip. $ -
   f) Total Acq.Exp.(c+d+e) $4,825
   g) Service Life   15 yrs.

Notes: Present Machine worn out - must be replaced.
$4,594 included in capital is tooling.

---

4. a) Net Capital Investment (3b - 2b) $57,507
   b) Net Cash Expenditure-after tax(3b + .48x3f + .25 x 2d - 2c - 2e) $59,335

5. Results Expected:
   a) Return on Net Capital Investment - Average - before tax   42.8 %
   b) "   "   "   "   "   "   after tax   20.5 %
   c) Pay-back - before depreciation - before tax   4.6 Yrs.
   d) Investor's rate method-after tax-interest rate on Net Cash Expend.   11.0 %
   e) Inventory change (decrease or increase - Cross Out One) $27,000 in 1960

Notes: Total potential inventory reduction for complete pole line is $44,500 @
$27.0 million operating rate. 60% of reduction is credited to This machine
& is expected to be accomplished. Remainder depends upon improved
assembly method.
Depreciation equivalent of alternate (ot replacement with machines of same
type as present) is taken as a 'cost saving' for proposed machine.

---

6. Important Intangible Considerations:

   Upon completion of line for pole making, factory order paper can
   be eliminated, and poles be produced in lot sizes of 3 sets or
   one days usage, whichever is greater.

Form 1903A2

FIGURE 1. Work Sheet I.

Capital Expenditure Analysis
SAVINGS AND COST DATA - BEFORE TAX

WORK SHEET II

Net Capital Investment (4a) $ 57,507

Operate Savings Net:
1st Yr. = Oper. Sav. Gross - Total Acq. Exp. (3f) - Depr. Net
Rest = Oper. Sav. Gross - Depr. Net

Year	Assignable Labor	Assignable Mat'l.	Rebuild or Retools Avoided	Drillings Avoided	Vise. Maint.	Inventory	Depreciation 1st replace	Depreciation 2nd replace	Install replace	Operate Savings Gross	Depreciation Net	Book Value Net	Operate Savings Net
1	5,440		4,750		900	3,470	1,010		500	16,100	11,460	44,050	4,640
2	6,040					3,860	900	1,040	500	13,240	5,970	40,080	7,270
3	3,930					2,510	780	900		9,020	5,190	34,890	2,830
4	5,440					3,470	690	780		11,280	4,550	30,340	6,730
5	6,550		4,750			4,190	600	690		12,930	3,970	26,370	8,960
6	5,750					3,470	510	600		16,180	3,380	22,990	12,800
7			300				460	510		11,730	3,010	19,980	8,720
8					(500)		390	460		10,970	2,590	17,390	8,380
9							340	390		11,360	2,220	15,170	9,130
10			300				300	340		11,260	2,010	13,160	9,250
11			4,750				260	300		15,630	1,690	11,470	13,940
12			300				220	260		11,100	1,480	9,990	9,620
13							190	220		11,030	1,270	8,720	9,760
14							180	190		10,990	1,160	7,560	9,830
15	5,760		300		900	3,670	140	180		10,940	950	6,610	9,990
16													
17													
18													
19													
20													
Total										IIa 183,716		310,770	IIa 132,860
Aver.												IIc 29,720	IIb 8,860

Results Expected:
5a. Return on Net Capital Investment - after depreciation - before tax (IIb ÷ IIc) = 42.8 %
5b. Return on Net Capital Investment - after depreciation - after tax (.48 x 5a) = 20.5 %

Note: If capital gain or loss is appreciable, add 75% of gain or subtract 100% of loss to Operate Savings after tax (.48 x IIa), then average and ÷ by IIc.

5c. Pay-back - before depreciation - before tax - = 4.6 Yrs.
(add Operate Savings gross to equal Net Cap. Invest. (4a))

FIGURE 2. Work Sheet II.

Capital Expenditure Analysis   Work Sheet III

Investor's Rate Method - Data and Calculation - After Tax

Net Cash Expenditures (4b) $ _59,335_
Net Capital Investment for
Depreciation Tax Credit (4a) $ _57,507_

Year	Operate Savings .48x11d	Deprec. Tax Credit	Net Cash Return	Int. 10% Present Value Factor	Present Value of Cash Return	Int. 11% Present Value Factor	Present Value of Cash Return	Int Present Value Factor
1	7,730	5,800	13,530	.909	12,300	.901	12,190	
2	6,360	3,120	9,480	.826	7,830	.812	7,200	
3	4,330	2,700	7,030	.751	5,280	.731	5,140	
4	5,410	2,330	7,740	.683	5,290	.659	5,100	
5	6,210	2,060	8,270	.621	5,140	.593	4,910	
6	7,770	1,800	9,570	.565	5,400	.535	5,120	
7	5,630	1,530	7,160	.513	3,680	.482	3,450	
8	5,270	1,320	6,590	.467	3,070	.434	2,860	
9	5,450	1,160	6,610	.424	2,800	.391	2,580	
10	5,400	1,010	6,410	.386	2,470	.352	2,260	
11	7,500	900	8,400	.350	2,940	.317	2,660	
12	5,330	790	6,120	.319	1,950	.286	1,750	
13	5,290	690	5,980	.290	1,730	.258	1,540	
14	5,280	580	5,860	.263	1,540	.232	1,360	
15	5,250	530	5,780	.239	1,380	.209	1,210	
16								
17								
18								
19								
20								
Total					62,800		59,330	

Results Expected:

5d - Investor's Rate Method - after tax - Interest Rate _11.0_ %

Form 1903C2

FIGURE 3. Work Sheet III.

Service life is the estimated time, in years, that the proposed facilities could be expected to be usable at the production volume planned, even though technological improvements could obsolete the equipment before it becomes worn out. In estimating service life, the following factors should be considered:

1. Number of shifts per day the equipment would be operated.

2. Severity of operating conditions (load, shock, vibration, heat, abrasives involved, corrosives present, building or mounting, conditions, etc.).

3. Equipment construction features to protect against severe operating conditions.

4. Skill of operators, setup personnel, and maintenance personnel.

5. Effect of turnover of operators and setup personnel.
6. Effect of incentives on equipment care, abuse, and maintenance.

If the product line for which the facilities are being proposed has a life shorter than the facilities, and will not likely be replaced by a product line that could effectively use the facilities, the service life should be shortened to equal the life of the product line, and salvage value of the facilities should be estimated and used as a saving in the terminal year. If *known* improved equipment, or new automated methods will become available in the near future, the service life for the facilities being considered should be appropriately shortened, and salvage value estimated and used as a saving in the terminal year.

SECONDARY MEASURES OF WORTH

Return on net capital investment, after federal income tax, is a secondary index of the worth of a proposed investment. Assuming a current 52 percent tax rate — except for proposals involving appreciable capital gain or loss on facilities that will be sold as a result of the proposal — the return will be 48 percent of the return on net capital investment before tax. If a capital gain would be appreciable, 75 percent of the gain should be added to the net operating savings total for the service life of the facilities, then 48 percent of that total used to compute the return on investment. (Twenty-five percent of the capital gain is federal tax.) If a capital loss would be appreciable, 100 percent of the loss should be deducted from the net operating savings total for the service life of the facilities, then 48 percent of that total used to compute the return on investment. (Capital losses can generally be deducted as business expenses before federal tax.)

Payback is another index of worth of an investment. It is one of the oldest measures used, and probably one of the least valuable. As years go by, in the opinion of this writer, it will probably be completely dropped from use by executives in the manufacturing and processing industries.

Payback may be calculated either before or after depreciation, and before or after federal income tax. Simplest, and most normally used, is the before depreciation, before tax basis. This gives an indication of how many years would be required to get the money required by the investment back into working capital. It is a *measure of risk*, but of itself gives no indication of the profitability of the proposal. It can be used to compare relative risks between two or more proposals, but cannot indicate which proposal would be better over its life.

INVESTOR'S RATE METHOD

The Investor's Rate Method is another index of the worth of an investment that is gaining in use. The Investor's Rate Method expresses a return on the total net cash expenditure, after tax, required by the proposal, in the form of an annually compounded interest rate. Calculation of this interest rate is a fairly complicated and time-consuming process. In theory, a proposed investment that would show an interest rate in excess of the interest rate that would have to be paid out to borrow money for the investment would be a profitable investment. This is because the Investor's Rate Method discounts future savings to a present value, and therefore fully takes into account the time-shape of the operating savings after tax.

As a practical matter, however, it is necessary to add reasonable safety factors to offset errors in the analysis data that exist in the various areas of forecasts, proposals, quotations, estimates, and guesstimates involved. Until such time as more background is available in the use of this index, an interest rate bogey for the Investor's Rate Method of the order of 9 percent is suggested for production automation proposed for cost reduction.

In this writer's company the Investor's Rate Method is used only under the following conditions:

1. When return on net capital investment is in the 30 to 50 percent range, the total acquisition expense is an appreciable proportion (15 percent or more) of the net capital investment, and the operating savings net in the first three to five years are low in relation to savings in the latter years.

2. When leasing is an alternative in financing an investment being considered.

EVALUATION POLICIES

In applying the various measures of investment worth to proposals covering automated equipment the question can arise as to whether there would be any difference as compared to considerations for ordinary equipment. No distinction should exist in the measures, or in the methods of reporting. Likewise, no distinction should exist in the compilation of data, calculation of returns, or in the policies controlling the derivation of costs and savings.

More careful consideration of many items is deserved, whenever expensive automated equipment is being considered, than is necessary when simple standard equipment is being considered. Such items include engineering costs, installation costs, debugging costs, setup costs, labor savings (both direct and indirect), material savings, maintenance costs, scrap and salvage savings, inspection savings, flexibility costs (rebuilding that might be required by product-line changes), floor space requirements, terminal values in cases where facilities may not be worn out at time of disposal, quality improvement, manufacturing cycle reduction, and inventory savings.

Intangible considerations must *never* be ignored, but whenever practicable should be converted to dollars. For example, quality improvement that reduces scrap should be evaluated in dollars, while quality improvement that is purported to increase sales should not be evaluated. Decreased delivery cycles that reduce production planning, dispatching, or inventory should be evaluated, while reduced delivery cycles that may increase sales should not be evaluated. Improved safety should not be evaluated.

One psychological reason for not trying to convert all intangibles into dollar evaluations is that opening of the door to "thumb sucked" dollar evaluations in the intangible areas may invite "thumb sucked" dollar evaluations in other areas of an analysis where digging and study would otherwise produce factual dollar values.

Important policies in the derivation of various items of costs and savings are:

1. *Labor*. Labor should be evaluated at straight time average earnings, plus fringe benefit cost, of the specific personnel involved when actual hours worked are used. Job rates, plus fringe benefit cost, of the specific personnel involved should be used when standard hours are used.

Overtime and shift bonuses should only be evaluated if directly affected by the proposal, and only to the extent affected, e.g., first year projected operating volume might require 2-shift operation plus five hours per shift overtime on existing equipment. and 1-1/2-shift operation without overtime for the proposed automated equipment. For the first year, shift bonus saving for 1/2 shift, plus overtime bonus of ten hours per week should be credited to the proposed equipment. Second year projected volume might require three-shift operation without overtime for the existing equipment, and two-shift operations without overtime for the proposed equipment. For the second year, shift bonus saving should be credited to the proposed equipment.

Indirect labor that may be specifically identified as a greater or lesser requirement of the proposal as compared to the existing requirement may be evaluated as under the first and second points. Indirect labor as a percentage of direct labor, or as an average of a plant or department, should *not* be used.

2. *Material*. Material should be evaluted at actual cost, not at standard cost.

3. *Overhead*. Overheads in terms of percentages of direct labor. or averages of plant or department, or fixed rates made up of variable expenses for a plant or department, should not be used. Only specifically identifiable items that would be affected by the proposal should be evaluated. Some of these items are power, perishable tools, supplies, supervision, and clerical requirements.

4. *Rebuilding*. Rebuilding of existing equipment should be considered as deferred maintenance, and charged as a cost in the year in which the work is to be performed, when such rebuilding is restoration from wear and tear rather than from basic design changes.

Rebuilding of equipment for basic design change, whether for existing or proposed equipment, is a capital expenditure. The additional capital investment for the rebuilding should be included in the analysis in the year in which the work is to be performed. This should be done to properly reflect an increase or decrease in the net capital investment (thus in the depreciation charges) and in the calculated return on investment.

Expenses incident to a capital rebuilding job should be included as costs in the year incurred. Such expenses as disconnecting, moving, cleaning, painting, reconnecting, operator time losses, and extra costs involved in subcontracts to maintain production fall into this class.

Proper handling of rebuilding costs is particularly important with automated equipment. Product line changes are most likely to be the reason for capital rebuilding of automated equipment. It is unrealistic to "sluff off" rebuilding costs as a necessity of the product line change and thus throw the cost against the redesign, while still showing a high return on investment on the automated equipment as originally proposed. Whenever product line change can be predicted during the service life of the proposed equipment, rebuilding costs should be estimated and be included in the analysis of return on investment on the proposed equipment. The net result of this method of handling might well be the selection of different equipment that would have a greater return over its service life than equipment considered without regard to product line changes.

5. *Retooling*. Retooling incident to wear is generally an expense and should be included as a cost in the year incurred.

Retooling incident to product line redesign is normally chargeable as a cost of the redesign, but should be considered directly in the analysis of return on investment of the proposed equipment, for the same reasons cited with regard to rebuilding costs.

6. *Floor space*. Costs of floor space can generally be ignored because differentials are not often significant. If sufficient areas are released to permit assured productive use of the area, or if less space would be required in a new building that is part of the proposal, then credit should be allocated to the proposal by way of reducing the net capital investment by the depreciated book value of existing space released, or the cost of new space saved. If extra space is required by the proposal, the net capital investment should be increased accordingly.

7. *Debugging*. Setup, operator and service labor, maintenance manpower, material scrapped, part salvage, supervision, manufacturing engineering, perishable tool breakage, inspection, utilities, subcontract production, and retooling are some of the items that make up debugging costs. Such costs are expenses in the year in which they are incurred.

General tendency is to be overoptimistic in estimating the costs of debugging equipment. Part of this tendency is due to inexperience and lack of accurate accumulation of all of the costs incurred. The remainder is an unwillingness to face up in advance to the potential costs known to be involved. An honest desire to recommend purchase of only those facilities that will show a true return on investment up to the established bogies, assisted by a postoperative audit of actual results vs. proposal results, can overcome the very human desires of manufacturing men to have the very latest, most highly automated systems available.

Clearly defined specifications of quality and performance included in the purchase agreement, with demonstration of results in the facilities manufacturer's plant before shipment or payment, will materially assist in minimizing debugging costs.

8. *Engineering*. Manufacturing engineering costs involved in the studies leading up to purchase of new equipment must not be ignored, but should *not* be charged as a cost in the analysis of return on investment.

Methods refinement, time study, standards establishment, incentive plan development, etc., involved after debugging may or may not be charged as a cost in the analysis of return on investment. This is a policy matter, but generally it is understood that these costs are not included in the analysis.

9. *Maintenance*. This is another area in which lack of information and overoptimism are quite often present. Training costs incurred in the upgrading of maintenance personnel in the intricacies of electronic, magnetic, and hydraulic components and their circuits should be considered as a normal cost of business, not something to be levied against any single piece of equipment. The labor, renewal parts (both used and inventoried), and supplies involved with the proposal should be analyzed carefully and should be included in the return calculation. Discussion with equipment manufacturers and users of similar equipment can provide reliable information to supplement the buyer's own experience.

10. *Inventory*. Costs of carrying inventory vary materially from one business to another, and are very elusive at best. Savings or extra costs involved in a proposal for new facilities should be conservatively estimated. Only specific inventory cost items affected by the proposal should be evaluated. These costs should be developed as a percentage of the dollar value of the inventory change resulting from the proposal.

One or more of the following items may be affected by the proposal: (1) cost of money (use the lower of the existing corporate capital distribution, or borrowing cost), (2) obsolescence (consider the particular product line involved), (3) material handling (estimate costs of handling into and out of the warehouse), (4) clerical and inventory taking, (5) taxes, and (6) insurance.

11. *Inflation*. No inflation or deflation should be considered. All figures used should be in current dollars. Caution is required to avoid inclusion of inflation that could be included in sales forecasts.

12. *Salvage value.* Salvage value of the proposed equipment is not normally a factor in the analysis, since the proposed equipment is normally expected to be used for its full service life. Its actual value becomes a factor in the "next generation" proposal.

As previously noted, if known improved equipment or revised methods will be available in the near future, and the service life has been appropriately shortened, then the salvage value of the proposed equipment may be appreciable and should be estimated and included as a saving in the terminal year.

13. *Sales forecast.* Although recorded last in this listing of important policies in the analysis, the sales forecast is the single most important factor in the evaluation of all capital investment proposals. Use of a five-year forecast, by years, is recommended as the basis for all capital investment analyses. In general, this five-year forecast is the basis for all corporate planning, and therefore deserves development by specialists in market analysis, combined with study and adjustment by top operations executives.

The sales forecast will most likely be in total dollars, with some breakdown into product lines, but probably not into specific product sizes or part numbers. The production planning department should convert the sales forecast into the product size and part number breakdown, making sure with the engineering department that planned product line redesigns are taken into account. Savings and costs, then, are developed on the volume figures from this breakdown of the sales forecast, by years, for the first five years of the analysis.

For the remaining years of the service life of the proposed facilities the highest "normal" yearly operating rate during the first five-year period should be used, unless other known factors (e.g., obsoleting of a line of product) would indicate some lower forecast for the remaining years. Selection of the "normal" rather than the highest yearly rate or the fifth-year rate is suggested to eliminate cyclical changes. It is unwise to follow a trend line for more than five years into the future, and gamble capital investment on such nebulous future expansion. The crystal ball used by all forecasters is cloudy enough when seeking answers for the next quarter, and is hardly even transluscent when looking five years ahead. Beyond that it is about as clear as the proverbial eight-ball when considered in the light of requirements for a particular proposed facility.

PAYOFF IN GREATER PROFITS

Simple forms suffice for recording the pertinent data for any facilities proposal. These forms assist in calculating the various indexes of worth of the proposal and summarize them so that operations executives can readily assimilate the important factors in the proposal. Back-up data should be available to anyone wishing to dig more deeply into the details of costs. savings, etc.

Use of the policies and measures of worth suggested, combined with accurately developed costs and savings applied to the operating levels predicted in a good sales forecast, should result in the making of wise, profitable investments. Profit potentials will be maximized, however, only if capital investment proposals for automated equipment contemplate the right degree of automation. Too little automation can reduce the potential return on investment, just as surely as too much automation.

Sound business judgment may be warped by the intrigue of automation. Be sure to confirm any judgment on a major expenditure for automated equipment by analyzing not only the return on investment to be obtained on the equipment proposal believed to be desirable, but also on equipment that is more highly automated, and on equipment that is less highly automated. In some instances a better investment than first thought possible can be realized by either spending more for more highly automated facilities, or by spending less for less highly automated facilities.

CONCLUSION

In any case, getting a grip on the real costs and savings involved in various proposals will pay off in greater returns on the investments involved, and therefore in greater profits to the enterprise.

In the final analysis, the *real* return on investment in automatic equipment for manufacturing may hinge on the financial policy of the company. Where little or no intensive study of the financial aspects has been made and no corporate policy established to cover all the factors discussed in this chapter, ineffective or nonproductive activity will be the result.

Without a firm perceptive financial policy, worthwhile and profitable long-range decisions can hardly be expected. It is unrealistic to expect profitable automation to "spring fully grown" from little more than sudden necessity for improved productivity. A plan and program are needed in order that answers are available when they are needed.

Not to be overlooked as an important part of the policy is the IRS Tax Code specifications in force. The policy must include selection of the basic code electives available. Serious study of the code requirements as well as the code recommendations will provide the best path to maximize return on investment in automation systems over a protracted period of time.

Consistency of evaluation is the true key to priority of investment.

REFERENCES

Management Problems in the Acquisition of Special Automatic Equipment by Powell Niland, Division of Research, Harvard School of Business, Boston, 1961.

Machine Replacement for the Shop Manager by Baxter T. Fullerton, Huebner Publications, Inc., Cleveland, 1961.

"A Better Way to Buy Any Capital Equipment," *Steel* Magazine, June 1, 1964, p. 35.

Industry Case Examples

CASE 8A

Automated Materials Handling Pays off for Small Molder

The question of the level of throughput at which automated materials handling becomes economical is frequently discussed at great length. And the reason for that is usually an accompanying shortage of facts. Very few companies, especially of the smaller size (which is really where the question of automation is most unresolved), develop the kind of data that would provide an answer to the question.

Here is one that did, and the facts collected make it quite clear that even if throughput is at only 1/2 million lb/year, automating is not only economical and convenient, it's profitable.

The company is Meyer Dairy Inc., Basehor, Kansas. It blow molds high density polyethylene milk bottles inplant, and started the operation in June 1969 at a rate of about 500,000 lb./yr. Previously the company sold its milk in PE-coated cardboard containers.

Obviously, such a small operation requires only the simplest materials handling equipment. In this case, it consisted principally of a 9 by 32 ft. epoxy-coated silo of 60,000-lb. capacity with a 45-deg. hopper; a conveyor to move the material from the silo into the plant; a couple of tote bins; and a vacuum machine loader to convey the material to the blow molding machine. Total cost of this system, including installation, came to $8000.

The company didn't have the silo installed until about three months after the molding operation began. During that time, material was purchased in 20,000-lb. quantities and delivered in 50-lb. bags. The company kept track of the labor cost of moving these bags to the machines: it was more than $3000 per annum.

In addition to savings in direct labor, there were savings in materials cost — both direct and through a reduction in contamination (pieces of paper and strings from the bags) — amounting to about $7500 per year. Storage-space savings came to $3500. Finally, there was an increase in productive capacity, resulting from the elimination of contaminated material. Production was raised from 1050 to 1200 bottles/hr. With each container worth about 6½ cents, this increased the value of finished product by about $9.75/hr., or over $20,000/yr. The detailed savings are summarized in the table (in Figure 4).

171

Material is now delivered to the dairy by truck in 40,000-lb. quantities. A blower on the truck loads the material directly into the silo; from there it is conveyed to the machine.

The storage and conveying system was planned and installed by Butler Mfg. Co., Kansas City, Missouri.

CREDITS: Reprinted by permission of *Modern Plastics Magazine*, McGraw-Hill, Inc.

Savings on labor:

Elimination of labor to unload 400 bags from truck and transfer to storage room (2 men, 1/2 day each, every 2 weeks). . . . . $ 936

Elimination of labor to move bags from storage room to machine hopper, open, empty and dispose of bags (1 man, 2 hrs. daily) . . . . $ 2,376

Savings on material:

Elimination of losses due to spillage and contamination (3% of 500,000 lb. at then current price of 17½¢ / lb.) . . . . $ 2,625

Through bulk purchasing (savings of 1¢ / lb. on 500,000 lb.) . . . . $ 5,000

Savings on floor space:

Storage space converted to additional refrigeration. . . . . $ 3,500

Gross annual savings:. $14,437

Less cost of materials handling equipment (installed) . . . . $ 8,000

Net annual savings . . . . $ 6,437

Value of additional production:

150 more jugs per hour (316,800 more per year, at no additional operating cost) $20,592

Total "dollars ahead" first year $27,029

FIGURE 4. How automated plastics handling paid off.

CASE 8B

Automated Sheet and Plate Storage

A $50,000 addition to Reliance Steel & Aluminum Company's first automated "pigeonhole" storage system for metal, credited with reducing material handling costs by more than 50 percent since it was installed nine years ago, now provides 570 new compartments, or pigeonholes, increasing the number of storage openings by 30 percent to a total of 2,460.

Each compartment accommodates up to 5,000 pounds of flat precut aluminum, magnesium, high-finish stainless or galvanized steel stored on pallets, (Figure 5) the metal staying there undisturbed until wanted. The racks, 22½ feet high, are served by two automated retrievers operating in parallel aisles (see Figure 6).

William T. Gimbel, president of Los Angeles-based Reliance, said the system has completely repaid its original investment in operational savings since its installation in 1962 at a cost of approximately $500,000. Net tangible savings to date are estimated by Mr. Gimbel at a total of $687,000.00

With capacity now increased by nearly one-third, he expects its money-saving potential to rise accordingly. Intangible savings are significant, though not easy to put into dollar values.

Contributing mainly to a reduction of more than 50 percent in handling costs, Gimbel said, were: (1) savings in manpower, (2) virtual elimination of material damage, (3) a sharp drop in personal injuries, (4) use of once-wasted cube space.

"The entire system can be operated by one man who punches 'instructions' to each retriever on a push-button console," he said. "This is the only manual operation.

"The retrievers can store one load and bring back another on one round trip, each trip averaging six minutes (see Figure 7). There is no lost time or motion.

"Beyond that, we have effectively eliminated the high cost of material damage which was considered a normal part of the business when metal was stacked on the floor and shuffled around by overhead cranes as wanted loads were searched out.

"Hurt hands, strained backs and other occupational injuries have also been reduced almost to the vanishing point." One major money-saver, he said, was a new concept the system introduced — capitalizing on what once was considered useless vertical space.

"The original racks gave us an effective 20,000 square feet of new storage space," Gimbel said. "The new racks add another 5,000 square feet.

"Based on 1962 values, 20,000 square feet of prime industrial space would have cost around $1,200 a month on lease or about $140,000 if we had bought land and put up a new building. It would probably be half again as much at today's prices."

The system is made up of four parallel racks. With the addition, one pair of racks has been extended to 300 feet, the other pair to 220 feet.

The original installation, including automated retrievers, was custom built by the Triax Company, Cleveland (see Figure 8).

CREDITS: Reliance Steel & Aluminum Company.

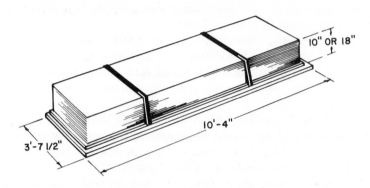

FIGURE 5.   Typical plate stack.

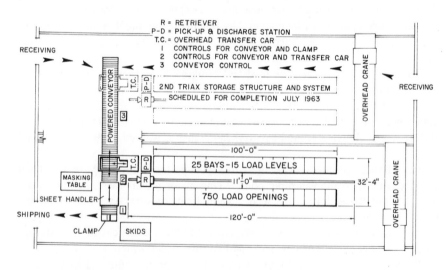

FIGURE 6.   Floor layout for automated sheet and plate storage system.

FIGURE 7.   View into stacker working racks.

FIGURE 8. No hands needed. Automated retriever in Reliance Steel and Aluminum Company's Los Angeles metal service center delivers load of palletized steel to start filling storage system's newly added racks. The $50,000 addition provides 570 new pigeonholes, making a total of 2,460 protected compartments. Company reports system, originally installed in 1962, has repaid entire $500,000 cost by lowering manpower needs, using once-wasted vertical space, other economies. (Photograph from Reliance Steel and Aluminum Company, Glendale, California.)

9

The Product Policy

Automation includes fully continuous operations as well as automatic step or batch-type operations that are fully or partly integrated. Which type is most suitable depends not only upon the product and process characteristics, but also on the conditions dictated by the state of advancement in the operations applicable.

OPERATIONAL STUDY

What is the fundamental effect of product design on automation in manufacturing? Can a basic study reveal significant factors for practical economic guidance? The answer is that product design characteristics can obliterate the value of automation unless serious thought is given to it.

While at this stage of evolution no simple engineering solution is available, it is possible to present some general rules for evaluation and study. Once the problem is reduced to finite proportions, work can be channeled into those areas which promise the most effective results. What is being sought is adaptability of the product to permit consolidation of specific processing steps into an economical integrated system.

It is fundamental to understand the primary reasons which can create the need to automate regardless of scale of operations or their general character. To assume or conclude that elimination of labor or reduction of usual labor is the main objective is erroneous. True, cases are on record revealing this as the primary purpose, but the majority indicate that the really important economic factors can be otherwise.

Need for increased productivity is the single major reason for the trend to automation. Other complementary reasons are present and often one of these alone *can* make automation a necessity.

Among the important spurs to automate is the process which cannot suitably be carried out by manual means of control. Here without automatic operation there can be no saleable product. Speed of the process is beyond man's capability to react. Linked with this, and again often the case today, quality may be the motivating force. Where the product must be of extremely uniform, controlled characteristics, most often only automatic control and/or production provides the necessary results. In addition must be those factors of safety hazards, undesirable working conditions, and monotonous or brute labor, no longer acceptable.

These factors then reduce to increased quality and uniformity, speed, safety, output (here reduced losses, waste, and scrap are important) and job attractiveness (here higher skills and wages become imperative).

TWO MAIN PHASES

In evaluating automation possibilities for any particular manufacturing process, there are two areas which should be carefully examined. These relate to the production method or methods and the automation system. These are generally approached in this order and only in special cases where an absolutely fixed product or process is concerned does the production method become inflexible. However, because of their tremendous influence on the cost suitability of the automatic system, product and process alterations should never be overlooked.

PRODUCTS AND METHODS

Significant importance must be accorded, therefore, to the product design and methods used when automating operations. Ingenious developments in both these areas have revealed the effectiveness of good product planning. Considerations that must be brought into the picture for successful automation are the detail design of the product, with specific concern for its adaptability to automation.

Both are critical. Without the first consideration, the product design may necessitate processing steps which are actually useless, thereby rendering the system more expensive or possibly impractical. Without the second, automation is often difficult or inadvisable.

Experience has shown that, regardless of the principles developed, the skill exercised in creating products capable of inexpensive and easy manufacture often spells the difference between success and failure. Many otherwise excellent products fail in the market owing to manufacturing problems. With automation this factor can be greatly magnified in its overall importance. And, in many instances, the important adaptation leading to success may appear relatively of insignificant proportions. Conversely, modern manufacturing techniques frequently remove inherent or old limitations on design that are often not recognized by the designer as deterrents to maximum design results.

DESIGN OF THE PRODUCT

The character of the product is always important. Few products exist that cannot be improved or simplified. Manufactured products such as hardware and mechanical items are especially critical in this regard. Final design aspects should never be "frozen" until the production requirements have been evaluated thoroughly.

In general, a number of specific recommendations have been made based on intensive study of the problem over the years.[1] No single recommendation can be set as *the* goal; each must be applied carefully for best overall results.

Strive for general design specification simplicity. Develop the product to attain simplicity of physical and functional character.

Production problems and costs have a direct relation to complexity of the product and any efforts expended in reducing a product to its "lowest common denominator" will result in real dollar savings. This applies virtually to any product component, and assembly or subassembly. Actually, simplicity is a great deal more difficult to achieve than complexity, but production costs bear a dismal relation to complexity of the product specifications and efforts to simplify will be well rewarded.

Examine all materials critically. Select materials not only for suitability but for lowest cost and availability as well. Seek economy of materials usage.

Selection of materials to be used in any product can plan a key role in attaining lowest costs. Special materials may create a production problem regarding availability or uniformity. As production increases, the material form and use assume a commanding role in the unit product cost. Overall possibilities to reduce materials costs deserve careful study.

Where applicable, use the most liberal tolerances and the least refinement possible on components. Be sure that surface condition and accuracies specified are reasonable and in keeping with the processes to be used as well as the product and its functions.

Refinement of surface conditions and dimensions play an important part in the final achievement of practical production design. It is imperative that the guiding principles of interchangeable manufacture be observed for successful low-cost production. All components of products should be examined to assure not only successful processing, but also rapid, easy assembly and maintenance.

Here, it is important to keep in mind the influence on practical production costs of standard available materials. Tolerances on dimension and geometry characteristics must be observed to assure suitability for automation.

Consideration should always be given to the possibility of automatic inspection. Where processing requires in-process inspection it is often desirable to minimize the number of critical dimensions subject to close control during processing. Automatic inspection will aid immeasurably in eliminating production losses and rejects.

Standardize as much as possible. Use standard available materials and parts and standardize basic components for more than one product to afford the volume needed for effective automation.

In developing products, use of standard components needs little emphasis. Standard stock items cost less and require no development. Also, in seeking best production design, standardization of components across product lines offers tremendous savings. In manufactured products this is a fertile field for attaining lowest costs and maximum output. Study of this phase has revealed some impressive opportunities and, in many cases, shown improved functional characteristics as a result of closely controlled automatic production made possible by the increased volume attainable.

DESIGN FOR AUTOMATION

Evolving the concept for a product is claimed to consume about 5 percent of the total effort needed to bring it to market; some 20 percent of the effort is required for design of the product, and about 50 percent to bring about economic production of the product. The remainder is usually involved with "debugging," design, production, and servicing. The big job today is reducing current costs; the greatest advantage afforded is through automated production.

Today, equipment design for successful automation is recognized to a much greater extent than is product design. Designers of components must be much better educated in the ways that automation operates. The designer who is familiar with automation and who designs his product with automated production in mind can make the work of system designers easier and can save his company thousands and even millions of dollars. The simple precaution of designing from the standpoint of ultimate automatic assembly can establish an automated operation rather than a manual one.

While it is possible to produce almost any product conceived, it does not follow that production costs will be reasonable or acceptable. Inflexible design centers on the product and its function, often creating entirely manual production requirements. The practical aspects of design modified for suitable automated production facilities should be carefully examined.

Research the best methods or series of fabricating steps. Then, redesign the product as needed to be adaptable to these methods selected as much as possible.

If innovation is carefully nurtured, it will be found that it is possible to have characteristically expensive parts which are inexpensive to produce via automation. Small but extremely important details can create this difference as can close observation of inherent limitations imposed by necessary automated processing procedures. For instance, the normal level of precision which can be maintained in continuous production must be observed to avoid high costs. Actually, for most processes, no amount of inspection will improve this level, but automatic control can produce the highest uniform quality.

The practice of specifying tolerances according the "rule of thumb" or empirical standards should be abandoned. Practical design for automatic production demands careful observance of the "natural" tolerances available with specific processes.

The most economic process will be influenced directly by the required quantity output. Many processes have quantity limitations and parts can be tailored to permit production by the method offering optimum cost advantage for the quantity and output speed needed. In process selection, the ideal material, acceptable costs and suitability for method of processing must be worked out in a cost/quality compromise. Material selection should be carefully made to insure maximum machineability, workability, castability, moldability, formability, weldability, or assembly, for most economical processing conditions and suitable results.

Evaluate the production steps needed. Plan for the most economic number of separate operations in manufacturing the product.

In automating, elimination of needless processing steps is imperative. Even though processing is automatic, useless production steps require time and expensive equipment. We find few products that cannot be improved in this regard. Small deviations must be carefully weighed for their effect on processing. Needless operations always create unnecessary handling operations and added cost. There can be exceptions to this rule where changes in the entire production sequence offer other possible savings.

Be sure to make handling simple. The aim is to simplify locating, setting up, orienting, feeding, chuting, holding, and transferring during production. A cardinal rule in automating is to complete making the product before releasing it, once processing is begun.

Automated operation demands that products be designed for most practical handling. With a goal of easy transporting of parts, hopper feeding, locating for automatic assembly, production cost is minimized. Mechanical products should be as rigid and compact as possible to withstand mechanical handling. Fragile products create production difficulty and necessitate complex facilities.

Discrete components should be capable of automatic feeding so that tangling is avoided. A good goal to remember is modular design amenable to automatic assembly. Avoid uniform symmetrical pieces that pose an orientation problem requiring expensive means for sure-fire handling.

An almost unlimited variety of handling devices is available but the simplest system always effectively reduces overall costs.

AUTOMATIC SYSTEMS

The important consideration that has been emphasized in the foregoing section is the key importance of the proper productive sequence for economic automation. Such equipment can range from a single multiple-stage setup to a complete production sequence. The line can be of simple open-loop fixed control arrangement or include the higher level of in-process measurement and closed-loop feedback control. The product characteristics, processing requirements, and practical considerations generally dictate the level of automation desirable. In the realm of equipment control, many new and potentially useful control methods are in development stages. The range of control is tremendous, hence, selection is more and more difficult.

For practicability and economy, along with suitability, it is imperative that a minimum of control be employed for obvious reasons. Added complexity in no way enhances reliability and adds to the cost of production equipment. If the memory storage of simple electrical controls is adequate for a particular application, to contemplate use of a more complex memory device not only opens wide the door to operating problems, but is a sheer waste of equipment and cost. Again, to plan on use of numerical controls for an operation where simple adjustable patchboard control will do the same job is just poor engineering.

As a general rule it should be said that the product character or the processes employed will largely determine the basic type of automatic system used.

While a high level of automatic control is practical, the use of full feedback automatic control of processing simply may not always pay off. The problem is one of being able to set up exact measurable specifications for the product against which comparison can be continuously made to maintain the product output in close agreement. Here is where a real problem exists — measurement of exact specifications is not always feasible. Means in many cases have not yet been found.

The important point, here, is that sometimes sophisticated feedback in the process is not necessary for successful results. And in certain instances it is actually impractical as well as not needed. Simplicity must be the watchword.

MEASUREMENT

In seeking successful automation, therefore, regardless of the type of operation, good accurate means for measurement will be needed for control. To exercise control it is necessary to measure the results of the operation being performed.

Certain types of measuring and sensing devices are readily available. Such straightforward measurements as temperature, dimensions, humidity, light, speed, color, flow, volume, weight, resistance, conductance, and pressure can be measured closely at high speed. Simple direct measurements such as these make practical the closed-loop control of individual segments of a system, but not always the total system itself.

IMPROVED QUALITY

This leads to an important concluding point regarding better and more closely controlled quality by means of automatic operations. Few areas so important receive so little attention. Automatic control of quality invariably results in more economical production. Tremendous savings can be made.

While better quality is desirable and offers real advantage in customer satisfaction, its other values can also be significant. Direct savings in materials can be made. Close control of quality substitutes elimination of defective products for the mere sorting of good and bad products so as to create additional savings in materials and labor. Better products can create ever wider markets and as a primary responsibility of management this increasingly important phase demands continuing and careful assessment. In tomorrow's market it will hold a prominent place.

FOOTNOTE

1 *Production Processes – The Producibility Handbook* by R. W. Bolz, The Industrial Press, New York, 1971.

REFERENCE

"Reducing Total Quality Costs" by Robert W. Traver, *Automation* Magazine, February 1971, p. 38.

Industry Case Examples

CASE 9A

World's Largest Fully Automated Brass Rod Extrusion Press

Free-cutting brass rod and bar in larger sizes and longer lengths in greater volume with greater quality control and faster delivery than heretofore possible in the brass industry is now produced at the Bellefonte Works of Cerro Copper & Brass Company.

This significant metal product development is just one customer benefit resulting from the installation of a new horizontal-extrusion press, a spectacular new 5,500-ton facility – the world's largest (see Figure 1).

Grey W. Tressler, the works' vice president and general manager, described the mechanical and metallurgical product implications of the press as "exciting and far-reaching." The press will extrude larger brass billets into rod, bar, or hot coils at the rate of one ton per minute to double the works' brass rod extrusion capacity to offer more production capability than any other similar facility in the copper and brass industry.

Because of fully-automated control cycling, he said, each successive billet, extrusion, and rod product will be an exact duplicate of the one preceding it. "This means each billet will be extruded at the same speed, the same pressures, the same heat as the last one to give a more uniform wrought structure in the brass. Metallurgical soundness, grain structure, surface finish will be uniform from lot to lot."

More uniform brass rod will provide better performance when producing machined parts, particularly on automatic screw machines. Further improvement in the grain structure of Cerro brass rod and bar, made possible by the new press, will also enhance such machining operations as knurling and thread rolling.

Rounds of free-cutting brass can be produced up to 8-inch diameter. Hexagons, octagons, squares and rectangles will be produced in longer lengths of large sizes than now possible. Installation of the new brass rod extrusion complex is expected to provide a denser rod product for large diameters that should expand markets for producers of machined brass rod parts.

The Automated Press. Employing new concepts in materials and machines, the Cerro heavy press (Figure 2) completes a new science born of the old art of

185

brassmaking. This is the most completely automated press of its type ever built. Among its many new features are solid-state, static-logic controls with press speed tied into the solid-state circuitry for synchronization of all stations in the extrusion cycle. The cycle of production is programmed to include billet ejection from the billet heaters. Sophisticated controls developed specially for this facility enable an operator to control all functions of the new press by a single push-button.

Other unprecedented advances include automatic butt and dummy block separation, two automatic wire reels synchronized with press speed, new coil transport system with up-ender, plate-conveyor runout table with transfer push-offs for subsequent operations. Auxiliary equipment includes cooling beds and the most modern automatic cutoff saw equipment — first of its kind. Brass billets to feed the press emerge from the largest and most advanced billet heaters, pioneered in this mill.

In the fully-automated press complex, the solid-state static-logic control elements are transistorized circuit boards that operate the hydraulic valving (Figure 3) and synchronize all press functions from billet heating to the run-out table and double coiler. Known as solid-state devices because they are capable of controlling electronic circuits without mechanical motion, these solid-state boards replace conventional mechanical relays in most of the brass extrusion press controls. A single circuit board does the work of four mechanical units nearly twice as large.

Another outstanding feature of this versatile press is the rapid tooling change for various billet sizes. A rotary die holder of new design permits use of duplicate dies during production of volume orders.

An operator of the new Cerro press pushes a button. This touches off the entire fully-automated extrusion cycle — from the time the billet comes out of the furnace through extrusion, shell removal, and rod handling, until the time the next billet emerges from the furnace. He can control the speed, switch to semi-manual operation, cause the press to "push" brass faster — a ton a minute — than ever possible in the brass industry. In short, the Cerro pushbutton press is a technological fulfillment in the automation of brass rod extrusion.

CREDITS: Cerro Corporation.

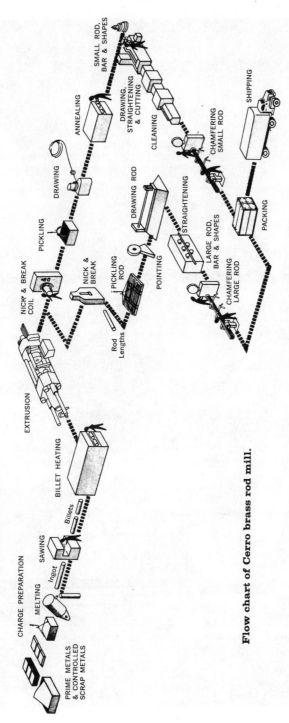

Flow chart of Cerro brass rod mill.

FIGURE 1. Flow chart of Cerro brass rod mill.

FIGURE 2. Close-up of Cerro heavy press shows ram (right) that applies up to 5,500 tons of power through dummy block to push the large hot brass billet through large die (center). Pressure exerted upon brass billet by the ram is intense, tremendous, absolute.

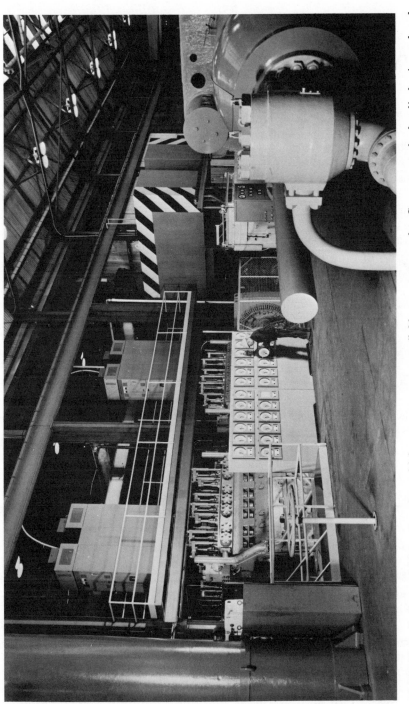

FIGURE 3. System which accumulates high-pressure water available to operate the Cerro water-hydraulic horizontal rod extrusion press (right) is shown here. Energy for the great press is stored in huge water and air bottles (left), largest ever built for a brass rod extrusion press.

CASE 9B

The Product Approach to Automation

To successfully achieve the required degree of automation, the assembly of product components must be considered at the onset and set forth as part of the specification for those components.

At AMP, Inc. corporate policy is based on the belief that excellence in product systems can only be achieved if the product to be applied, the device receiving this product, and the assembly equipment are considered as a *total* concept — each dependent on the other and all developed simultaneously. This in many cases will dictate that the user select a supplier on his reputation for excellence and integrity in the concept stages of his new development in order to obtain a satisfactory manufacturing concept.

Consideration of the total concept of components and assembly allows the supplier to offer the user a staged tooling approach. Where the ultimate system is devised in concept form in the initial stages, the supplier can then develop the key elements in the initial stages of the program and offer the equipment user the various production capabilities required through his prototyping, production sampling, initial production, and finally full scale production. With this approach the equipment user expends his capital equipment outlays only as his levels of productivity increases. This also aids the tremendous value of production conditioning, the ultimate systems that the manufacturer will use prior to full scale production demands.

Designed by AMP, Inc., and used for the manufacture of automotive brake safety switches, the machine shown in Figure 4 assembles *and tests* up to two thousand switches per hour. The eight component parts are fed automatically into predetermined locations and machine-checked for accuracy of location before the next operation is performed. When the switch has been completely assembled, the machine checks five critical performance features and accepts only those switches that meet customer requirements. Defective switches are rejected and scrapped. Because of this 100 percent testing feature and because of the switch design, AMP has produced and delivered over two million switches without a reported failure.

The close-up (Figure 5) of the automotive brake safety switch assembly machine shows the five inspection stations that test for free position tolerance of plunger height, minimum plunger height tolerance, maximum electrical contact tolerance, minimum electrical contact tolerance.

Failure of the switch to successfully pass any one of these tests results in automatic rejection by the machine. After testing, satisfactory switches receive an identifying mark and are dropped into a receiving hopper.

CREDITS: AMP, Incorporated.

FIGURE 4.   Automatic assembly machine for automotive brake safety switches.

FIGURE 5.   Close up view of safety switch testing operation.

CASE 9C

Automation in Electronics

In the words of Richard C. Oeler and Merlyn M. Armstrong of Motorola, Inc., "To make a television circuit module at a competitive cost, automated production is practically a must. Component and termination insertion must be adapted to automated assembly. Clinched leads eliminate trimming and improve solderability."

Using the four specially developed machines shown in Figure 6 one operator can apply over one-half million electrical contacts to over 25,000 circuit boards in one eight-hour shift. These machines are used by Zenith in the production of Dura Module circuit boards for their Chromacolor TV sets.

Serviceability. The customer's cost of maintenance and ease of service are synonomous. In light of today's critical shortage of competent servicemen, it is essential for television receivers to be maintained without the need for a high degree of technical knowledge and craftsmanlike skill. Modular packaging offers one approach to this problem. Such plug-in modules must be inexpensive enough so that any independent serviceman has, or has access to, a full complement. To accomplish this, the number of different module types must be minimized, and future modules should be interchangeable with previous ones. For example, 1975 modules should be directly interchangeable with those used in sets made in 1970, 1971, 1972, etc., thereby requiring the field service organization to stock only the latest units. Also, this would give the customer any extra performance bonus that the most up-to-date circuit design provides. These requirements serve to reinforce the demand for flexibility in the packaging concept. Ideally, such modules would be readily repairable by the service technician, because as desirable as printed circuit boards may be for modularization, there have been many objections (real or imagined) by service technicians to their use.

The 25-machine conveyor line shown in Figure 7 applies 30,000 electrical connections per hour to speed the production of printed circuit modules used in Motorola Quasar color TV sets. This line was a joint effort of engineers from Motorola and from AMP Incorporated, manufacturer of the interconnection devices used. Through the use of the specially designed interconnection devices, printed circuit boards become pluggable modules that speed up assembly and greatly simplify field servicing.

CREDITS: AMP Incorporated.

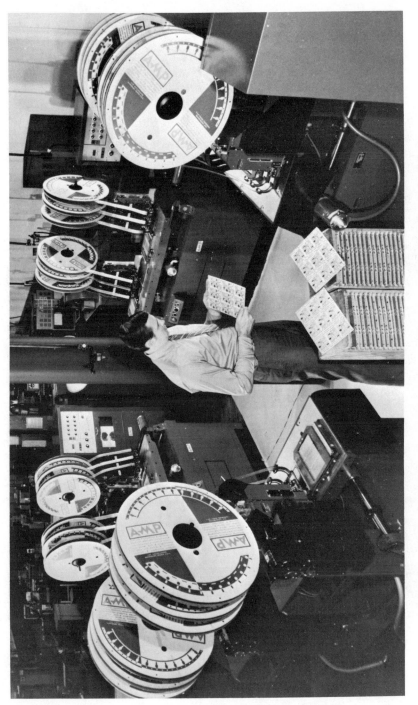

FIGURE 6. Flexible setup for insertion of electronic components into circuit boards.

FIGURE 7.   Continuous automatic assembly of printed circuit boards.

10

Manufacturing Engineering Policy

A good automation program calls for advanced manufacturing engineering with a wide range of talents to cooperate with and carry out the plans of the systems group. These talents are directed not so much toward the design of a product or to the supervision of workers who make it, but to the problem of *how* to make it. The men in manufacturing engineering must understand the product in all its ramifications, its area of use, and must be dedicated to the attainment of lowest costs. They should be familiar with many processes and how to design equipment systems and entire plants that make the processes possible. They should be able to integrate the product design and the manufacturing processes while recognizing the importance of the economics of manufacture. They should be represented on the systems group that determines the automation requirements.

Actually, the problems of product design are inseparable from those of manufacturing processes, and the intricacies of automation cannot be completely separated from those of plant management. Hence, good organization will keep in sight the interdependence of production, manufacturing engineering, and product design. Cognizance by manufacturing engineering of corporate financial facts is also of key importance.

To insure successful automation, it is necessary in good practice to define the overall functions and responsibilities of manufacturing engineering. Specifically, the responsibility for the design creation of a product should be that of the product design group. As a corollary, changes in the product character, whether they originate in or out of product group, should be made known to it and cleared by it. Similarly, since manufacturing engineers are responsible for automation, changes needed or brought about by the imperatives of automation should be fully discussed and explained by them. In the same way, instructions or training of workers required by the new automation should clear through plant supervision early in the automation program.

In fixing physical, electrical, chemical, appearance, and safety characteristics, careful consultation between the product people and manufacturing engineers will help insure that the product can be made at reasonable cost by known processes or by processes that can be economically developed. In order to make preliminary cost estimates, and arrive at a reasonable decision for action, it is usually necessary to agree or compromise as to the imperative attributes that can be accepted. This compromise between product specifications and a reasonable production cost result becomes increasingly important as the use of highly automated production equipment grows.

DEVELOPING THE PROJECT

The soundness of any automation project rests to a great degree on the care exercised in the handling of all the small items that go to make up the whole. A major project requires skill with economic studies, experience in introducing new parts and assemblies, and good judgment as to the time required to introduce them. Practice with a wide variety of manufacturing methods leads to quick selection of the most suitable processes and accurate appraisal of their economics and typical product quality. Experience with former projects yields orderly transitions from one product to another and freedom from unexpected delays in product schedules. Experience in bringing all the specialized skills needed to bear on the problem provides a degree of confidence in undertaking entirely new processes rather than patching up old methods.

In order to attain this degree of confidence, the manufacturing engineer must possess the ability to keep up to date with a variety of manufacturing technologies, reliable and knowledgeable technical consultants, his own internal operating organizations, and the corporate accounting organization. His work will be unsuccessful if he does not understand that manufacture of complex and high-quality products depends upon his detailed concepts and instructions for each step in manufacture and that his control over these instructions greatly facilitates initiating changes that inevitably arise.

Thus, organizing the attack on an automation problem is as important an element in achieving a successful solution as the final system design itself. The particular organization plan, or method of attack, will vary widely depending upon the character of the problem, the engineers available, the knowledge that can be brought to bear on the problem, and the time allotted for reaching a solution. However, there are certain common elements to be recognized in attacking any automation problem, and a *positive* plan of action should be created or vacillation will result in little but "shuffling paper." Key factors in determining the ultimate success of an automation project can be pinpointed as: selection of personnel to work on the problem, effective use of any available engineering talents on specific portions of the problem, and *a clear, methodical approach* to the problem.

DEFINE THE PROBLEM

To define the method of attack that should be used, it is necessary first to clearly define the automation problem to be solved. Of prime importance is a definition of what processes or operations are to be considered for automation. This factor cannot be overemphasized! Next, the objective of proposed automation should be clearly defined. Beware, here, as altogether too many projects in this writer's experience suffer from either a lack of clear-cut objectives or utterly confused objectives that change from day to day.

A final area of definition is the limits to be placed on studying and designing the *ultimate system*. In most cases the introduction of automation into an existing organization will affect many areas beyond the particular manufacturing area involved, Production control, purchasing, quality control, assembly, raw stores, or accounting may be affected by the introduction of a simple change to automation. How many of these areas has the systems designer taken into consideration? By specifically defining these limits, the manufacturing engineering organization can be given a relatively clear area within which to work.

An automation project, in addition, will usually require consideration of at least one typical problem that arises: what should be studied?

Thus, with the problem defined, it is possible to study and design the required automation system. The "task force" type of organization is probably one of the most effective methods of attack.

THE FEASIBILITY STUDY

As mentioned in Chapter 1, it is wise to raise the question: Do you really know where automation can make a real payoff? The answer was that you don't know, and can't know, unless a thorough "feasibility study" is made. This should cover all phases of the manufacturing system to pinpoint the steps to take, where, and in what sequence. Equipment policy should include such an approach and provide the means by which it can be done.

This policy includes study of the dollar returns on investment made possible by the contributions from new technology, from manufacturing research, and/or outside engineering expertise. Finally, it is in this area that the company must enforce a reasonable policy on acquisition of new capital equipment. Often, little or none of the potential payback is realized on equipment purchases made solely on the basis of the lowest bid.

Creativity — the heart and soul of automation — carries with it the powerful seeds of productivity, but these are never revealed in a dollar price. Best answers can be had only from competent firms that can work cooperatively with the buyer's manufacturing research group. Full knowledge and understanding of the production problem to be solved are necessary to attain the available profit potential, and this can show up *anywhere* in the total system.

Because the development of unique, automated systems of production is largely of research-development character, the most practical and cost-effective approach is in the light of total production needs. Such an approach involves what can best be termed an engineered master plan which attacks the basic problem by means of a phases program, each step of which leads directly to the next with maximum economy. The major steps that should be included are:

1. Preliminary engineering and production system survey. Analyze the problem. Study the state-of-the-art and automation technology applicable.

2. Concept development and system feasibility cost study. Test mockups and determine system economics.

3. Design and build programs. Complete development of details. Design, build, and debug equipment.

4. Study system refinement, Build, install, and debug system additions and improvements for added productive capacity, quality control, and computerization as required.

CONSIDER OVER-AUTOMATING

During the process selection and design phase, management should give consideration to "over-automating" as a means of increasing productivity on a long-range basis. This will probably provide more capacity than is necessary to meet the immediate production requirements. With this, some equipment may be idle for certain periods while the production requirements are being met, or lower costs could help market penetration. This situation is highly favorable if there is a remote possibility that market demand may be increased in the near future. However, the ultimate cost or quality of product must be weighed against costs of the additional equipment. This possibility can, for nominal expense, often contribute a sufficiently significant added value to the product to gamble on the possibility of ultimately using the facility at full capacity.

This consideration may be illustrated by the manufacturing system that enables inventories of finished assemblies to be kept at a low level. Thus, with a programmed assembly line, quick changeover time and complete flexibility of production are achieved. As a result, a stock of finished assemblies need not be inventoried if the physical location of the automated production line is in the area where finished assemblies are required.

CONCLUSIONS

As well chosen, as well grouped, and as well manned as any such broad organization of activities as those described may be, it is of little value unless it has the support of the other organizations in the enterprise. Informal acceptance and support by the other organizations must be based upon recognition that in their field, the manufacturing engineers are experts. The means for supporting the effectiveness of manufacturing engineers come first from giving the details of manufacturing processes that are issued by the engineers the force, not of suggestions, but of requirements of shop production, and second from assigning to manufacturing engineers control of expenditures for all new or changed manufacturing facilities. Although the first gives control over manufacturing processes and methods on the basis of assigned authority, it has little value without the second means. Without responsibility for expenditures for plant facilities, the best of engineering plans, the most careful of economic studies and the most active cost reduction program is paper work. When the engineer has to make a reality of his planning by providing the facilities he has specified and then comparing actual costs with his cost estimates, a measure of manufacturing engineering effectiveness becomes possible.

REFERENCES

The Manufacturing Man and His Job by Robert E. Finley and Henry R. Ziobro, American Management Association, New York, 1966.

Automating the Manufacturing Process by George F. Hawley, Reinhold Publishing Corp., New York, 1959.

Production Automation and Numerical Control by William C. Leone, The Ronald Press, New York, 1967.

The Manufacturing Engineer — Today and Tomorrow, a Report to American Society of Tool and Manufacturing Engineers, C-68922, July 1968.

Industry Case Examples

CASE 10A

Automation Solutions in Small Component Handling

The Application: Printed Circuit Boards (PCB) can be manufactured efficiently in large lots on automated machinery. After all of the electronic components have been positioned on the PC boards, they can be flow soldered on a moving production belt that permits each board to be completely soldered in a single pass.

The Problem. The savings gained by automated assembly can be reduced if hand insertion of components is required. The components are of different sizes and have different length leads, and they are inserted into locations all over the PCB. This also makes hand loading quite tedious. For a machine to insert components in a PCB, it must first find the correct component. It must then adjust itself to the proper configuration for holding the component. Next, it must deliver the component to the assigned location. Finally, the machine inserts the component, crimps the leads, and performs a check to see if all this has taken place in the proper sequence. If everything is operating properly, the machine releases the component and goes back for the next — which may be of a different size and shape.

Because of the many variables, using hard-wired logic to control the machine is very complex and expensive. Furthermore, if the machine is to be used to load other designs of PCBs, the hard-wired logic would have to be changed at considerable cost. This modification expense would occur each time a production run is made with a PCB design different than that of the previous run. To keep machine systems within practical limits of complexity, human operators perform some of the functions for loading — position location, crimp check, etc.

The Solution. To make the machine loading system automatic one manufacturer incorporated a General Automation SPC-12 as the machine controller (see Figure 1).

The computer controls five different drives on the machines: the X and Y movements of the PCB loading table, the drive on the component insertion fixture, the holding fixture, and the moveable stops that regulate the depth of

201

the insertion stroke. Target instructions for table movement are delivered by the computer program. This also keeps track of where the load table is and in what direction it is moving.

After insertion of the component, the computer makes an electrical continuity check to determine that the component is in its proper position and that the leads are being crimped. It also checks on the system to see if it detects an error programmed in the sequence.

Instructions are initiated by the SPC-12 and a feedback loop from the sensors determines that they have been properly executed.

A linear encoder is affixed to the table which provide the computers with table position to 0.0005 inch. By continually comparing the actual position with the target position, the computer generates an error signal. If an error signal is sufficiently great, the drive motor is given its full voltage. As the error decreases the drive voltage drops until it reaches zero as the target error reaches zero. An identical procedure drives the Y movement and the component size adjustment drive.

The System.   The component loading table's two degrees of freedom are on the horizontal plane. Separate driving motors move the table in the X and Y directions. The table moves under a component holder and inserter that is fed by a belt loader. Jaws on the component holder are adjusted and controlled by the SPC-12 to pick parts off the loading belt.

As the table is moved under the holder, a unique optical digital system determines its precise location to an accuracy of 0.0005 inch. Compressed air supplies the downward thrust to the component holder. One of several different moveable stops are pushed into place to limit the insertion stroke. As the component leads pass through the holes in the PCB, they come in contact with metal turning stops that bend and crimp the leads on the bottom side of the PCB. Electrical continuity between the component holder and the turning stops indicate that the leads have actually made contact. Having made contact, they will be bent; there is no other way they can go.

The computer program knows the size of each component as it is picked from the loading belt, and it causes the jaws' drive motor to adjust accordingly. The program also selects which of the insertion stops will be used, then it triggers the insertion drive at the proper time. If, during insertion, the computer fails to get the proper electrical continuity signal, it stops the procedure and sounds an alarm.

To self-check its own failure detection program, a blank space is deliberately left in the loading belt and the table is programmed to a dummy location. The machine goes through all of the operations. If it should get an electrical continuity indication for this position, the computer would know that something was wrong and stops the procedure.

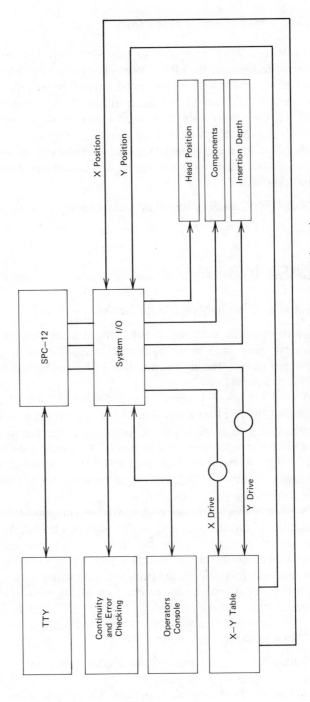

FIGURE 1. Computer controlled component insertion system.

The Value. When compared to machines controlled by standard numerical control techniques, the SPC-12 controlled system can insert components twice as fast. Up to 22 components can be inserted in one printed circuit board in 17 seconds. In addition to the increased speed and reliability provided by the SPC-12, the system is also more flexible as a completely different printed circuit board can be accommodated by merely loading another paper tape into the SPC-12 via the system paper tape reader. Production records, inventory control information, and performance logs are also available on demand at the request of the operator.

CREDITS: General Automation, Incorporated.

CASE 10B

Automatic Assembly of Door Locks

A project which calls for a well defined objective and properly organized manufacturing engineering approach is typified by a complex automatic assembly system. The automotive door lock assembly machine shown in Figure 2 is such a project.

Tooled for assembly and testing of front door locks, the machine is 28 feet wide and 107 feet long. Cycle time at any station does not exceed three seconds. Production rate is in excess of 1000 assemblies per hour consisting of 21 individual parts. Each machine — one for right-hand and one for left-hand locks — consists of ten manual stations and thirty automatic stations. Operations include automatic feeding, placing of components, welding, spinning, testing, and date stamping.

Nondestructive testing and probing stations continually check and monitor the condition of the assembly and, in the event of a malfunction or failure, trip the mechanical memory flag on the fixture carrier. Subsequent operations are discontinued on this carrier and the reject station is initiated which removes the defective assembly prior to the automatic date stamp station.

Each pallet is individually powered permitting unsynchronized independent transfer from station to station. In addition, modular station construction provides complete flexibility to accommodate future model or tooling changes.

Station sequence is as follows:

Station No. 1 — Automatically feed, orient, and place fork bolt stud including a one-cubic-foot supply hopper.

Station No. 2 — Automatic probe and manual backup station.

Station No. 3 — Automatically feed, orient and place two fork bolts. Parts to be fed from one bowl and shuttled into two pick-up positions. This station to be equipped with a magnetic belt type elevator bulk supply hopper to feed bowl on demand.

Station No. 4 — Automatic probe and manual backup station.

Station No. 5 — Automatically feed, orient, and place detent stud, including a one-cubic-foot supply hopper.

Station No. 6 — Automatic probe and manual backup.

Station No. 7 — Automatically feed, orient and place two detent levers. Parts to be fed from one bowl and shuttled into two pickup positions. This station to be equipped with a magnetic belt type elevator, bulk supply hopper to feed bowl on demand.

Station No. 8 — Automatic probe and manual backup station.

Station No. 9 — Automatically feed and place P/B stud including a one-cubic-foot supply hopper.

Station No. 10 — Automatic probe and manual backup.

Station No. 11 — Automatically feed and place transfer stud including a one-cubic-foot supply hopper.

Station No. 14 — Manually load P/B lever.

Station No. 15 — Manually load transfer lever.

Station No. 16 — Manually load intermittent lever. We are to retool a vibratory bowl supplied by the customer to feed and orient the intermittent lever and position for manual pickup.

Station No. 17 — Spare manual.

Station No. 18 — Automatically spin intermittent level stud. Station to include Paasche spray mist tool lubricators.

Station No. 19 — Automatically lubricate F/B and D/L.

Station No. 20 — Manually load frame.

Station No. 21 — Manually load frame.

Station No. 22 — Spare manual.

Station No. 23 — Standby spinner with X and Y positioning to cover F/B, D/L, P/B, transfer or intermittent studs. To include Paasche spray mist tool lubrication.

Station No. 24 — Automatically spin D/L stud. To include Paasche spray mist tool lubrication.

Station No. 25 — Automatically spin P/B stud. To include Paasche spray mist tool lubrication.

Station No. 26 — Automatically spin transfer lever stud. To include Paasche spray mist tool lubrication.

Station No. 27 — Automatically spin F/B stud. To include Paasche spray mist tool lubrication.

Station No. 28 — Manually turn over and inspect. Station to include an air-operated unloading assist device.

Station No. 29 — Automatically feed and place silencer. To include a one-cubic-foot supply hopper and a detent lever positioning device.

Station No. 30 — Manually assemble pin and spring.

Station No. 31 — Manually assemble pin and spring.

Station No. 32 — Manually assemble pin and spring.

Station No. 33 — Spare manual.

Station No. 34 — Automatically feed and place L/L stud. To include one-cubic-foot supply hopper.

Station No. 35 — Automatic probe and manual backup.

Station No. 36 — Automatically feed and place R/L stud. To include one-cubic-foot supply hopper.

Station No. 37 — Manually assemble L/L.

Station No. 38 — Manually assemble L/L.

Station No. 39 — Manually assemble R/L.

Station No. 40 — Manually reposition frame on fixture and position L/L.

Station No. 41 — Spare manual.

Station No. 42 — Automatically spin L/L rivet. To include Paasche spray mist tool lubrication.

Station No. 43 — Automatically spin R/L rivet. To include Paasche spray mist tool lubricating.

Station No. 44 — Standby spinner with X and Y adjustment to handle either L/L or R/L rivet. To include Paasche spray mist tool lubrication.

Station No. 45 — Spare manual.

Station No. 46 — Manually assemble over center spring.

Station No. 47 — Manually assemble over center spring.

Station No. 48 — Automatically clinch pin and push out test clinch.

Station No. 49 — Manually reposition lock assembly on fixture.

Station No. 50 — Manually assemble backplate to frame.

Station No. 51 — Spare manual or automatically probe for presence of two back plates. If two are not present, flag both fixtures so as to reject both assemblies.

Station No. 52 — Automatically hot upset back plate tabs.

Station No. 53 — Automatically hot upset back plate tabs.

Station No. 54 — Automatically hot upset back plate tabs.

Station No. 55 — Spare manual.

Station No. 56 — Manually assemble F/B spring.

Station No. 57 — Manually assemble F/B spring.

Station No. 58 — Automatically probe for correct assembly. Probes to distinguish L/L, P/B lever and R/L.

Station No. 59 — Automatically test and color code rejects two at a time. Identify failures with the color black.

Station No. 60 — Automatically test and black color code rejects two at a time.

Station No. 61 – Automatically eject bad assemblies and wrong models.
Station No. 62 – Automatically lot stamp good assemblies.
Station No. 63 – Manually unload completed assemblies.

CREDITS: Teledyne Precision, Cincinnati.

FIGURE 2.   Machine for automatic assembly of door locks.

CASE 10C

Automotive Project Management

The Vega assembly system is geared to high-volume efficiency because better productivity is essential to combat successfully the pressures of inflation and foreign competition. The Vega assembly line shown in Figures 3 - 9 also features the most advanced quality control procedures ever attempted.

All planning, from the start, was coordinated on an interdisciplinary basis. Extensive use of the project manager concept involved GM supplier divisions and key outside suppliers. This manufacturing engineering project concept is an important new trend in vehicle manufacturing.

The Vega developmental team consisted of engineers from the fields of design, process control, quality control, systems, reliability, and, of course, manufacturing. It also involved mathematicians, data processing specialists and material handling, parts and service experts.

This approach is especially important from the standpoint of lead time in bringing a product to market. Close coordination among staffs and organizations assure that all major design and manufacturing decisions are compatible with each discipline's specific responsibilities in the overall project.

New and more extensive computer applications helped to make the most of the lead time. One of these is the concept of variation simulation. As part of an extensive pretest program, three-dimensional mathematical models of individual components were made and analyzed by the computer. The objective was to determine if any variables could prevent the part from functioning as it was designed. These variables include such factors as tolerance limits, the materials used in the part, manufacturing processing and operator functions and sequencing.

In the vast majority of cases, simulation verified the accuracy of designs as submitted. Modifications were made as indicated. Simulation has been used before in selective ways, but this is the first time, in GM history, that it was applied to an entire vehicle. By applying simulation to the manufacture of critical vehicle systems, it was possible to assemble, in effect, thousands of Vegas before the car left the drawing board. Simulation helped to indicate changes in assembly line procedures and in the way operators make certain component installations. Job content for individual operators was a very important part of these considerations.

Planning was directed at making each car right the *first* time, in station, which helps to minimize the cost and time involved in out-of-station repairs before a vehicle is released for shipment.

Computer technology was also being used to support the Vega inspection system and to make each employee's job on the line more convenient to perform in a more precise way. One example is the ALPACA system — Assembly Line Production and Control Activity. With the aid of a central computer, better equalization of work loads for individual employees was made along the mile-long assembly line. ALPACA is designed to make sure that each employee has the right amount of time to perform assigned tasks in the right way. In essence, the system steadies the flow of production for greater overall efficiency, promotes better quality, and helps to improve employee morale and safety.

ALPACA anticipates and sorts out all of the variables that can make production line balancing an industrial nightmare. The system indicates to foreman of the floor through console monitors what effect normal changes — such as constantly changing model mixes or customer-ordered options — will have on individual work stations.

Another example of computer technology — one of the most discussed features of the Vega system — is known as PACS, an acronym for Product Assurance Control System. It encompasses two primary activities. The first involves developing process review standards for key assembly line operations. This is accomplished by a process review team of processing, industrial, and quality control engineers.

The team studies key operations to determine their capability to conform consistently to prescribed quality levels. Virtually every facet of these operations is evaluated from work space layout to employee training and orientation. When the review is completed, the operation is functioning at maximum efficiency — the perfect blending of men and their skills, material and machines — and everyone connected with it knows what must be done to keep it that way.

The second part of PACS involves a "closed-loop" reporting system. Its purpose is to assure that all the quality designed into the Vega reaches the customer. Preprinted inspection tickets accompany each vehicle from the body shop to the end of the assembly line. Inspectors along the way record their findings on these tickets simply by checking the appropriate code. Their reports are transmitted to a central computer through sixteen optical scanners located at strategic places in the assembly cycle.

At each stage, the computer compares the in-process quality level on all operations to the prescribed standards in its memory bank. Discrepancies result in an "alert" message which is transmitted to the floor according to the severity of the problem. The message is sent immediately via the teletype, in the case of serious discrepancies, to a process monitor in the station involved. He follows a checklist, structured by probability of cause, to identify the problem. Then he orders the necessary adjustments.

PACS is a "closed-loop" system. No serious unresolved problem is allowed to continue beyond a specified period. No problem goes undetected. In this way, PACS provides a real time rather than historical back-up for the Vega inspection system. It accounts for every critical inspection for each unit on the assembly line.

Before the car is okayed for shipment, which involves a recheck and reprocess inspection procedure, PACS performs its single most important function. The final inspector puts each car's serial number on a teletype to the computer. It instantly reviews the unit's assembly history and reports back to the inspector on the unit's readiness for shipment.

CREDITS: Chevrolet Motor Division, General Motors Corporation.

FIGURE 3.   The engine and rear-axle assembly are positioned by hydraulic lifts.

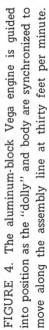

FIGURE 4. The aluminum-block Vega engine is guided into position as the "dolly" and body are synchronized to move along the assembly line at thirty feet per minute.

FIGURE 5. Wheels are put on the car by an operator using an air-wrench which tightens all the attaching nuts at the same time.

FIGURE 6.   An overhead conveyor with four elevations makes each installation along the line easier to perform, as this operator demonstrates by adding the plastic grille.

FIGURE 7.   The Vega touches the floor for the first time in the assembly procedure. Next installation is the bumper and valance.

FIGURE 9. Giant vents which remove exhaust straddle a Vega on road simulator that duplicates virtually any road condition.

FIGURE 8. Hatchbacks up and doors open, Vegas head down the final line where carpets and seats are installed.

11

Manufacturing R & D Policy

Are you investing for the future? In considering automation in this light, it is important to note that there is a direct relationship between the success of any manufacturing business and the inclination and ability of its managers to risk time and money in the prospects of creating improved operation from both a quality and cost standpoint.

The key is in the word "risk." When it comes to research in manufacturing processes and development of automated equipment, many industrial managers usually adopt a pretty difficult position — avoid risks, demand guarantees. Management seldom associates the basic concept of entrepreneurial risk-taking with such research and development to make the business more efficient. Yet, somehow, down-the-line supervisors and manufacturing engineers are expected to keep costs in line by some sure-fire means or other that involves little expense in learning the "how-to."

REAL CHALLENGE TO PROFITS

In this decade of rapid technological advancement, of changing markets and products, of intense competition both at home and abroad, there will be no easy

215

status quo. Business will require the will and ability to translate engineering knowledge and research into hardware for greater productivity. It takes no crystal ball to see that the real challenge and the real profits will lie in the practical application of the results of manufacturing R & D.

Both short and long-range manufacturing research must attain the same position of importance as product research. Industry has long been accustomed to taking calculated risks in the development of new products. Parallel manufacturing engineering research can pay even higher dividends; for the economics of production can open or close the door to financial success.

A newspaper production case in point is noteworthy:

"Chapman and Jim Knight have wrought similar economics and savings in the composing room. Today, every KNI paper uses a computer not only for administration and typesetting, but to control presses and other equipment, as well. That even includes KNI's *Boca Raton News*, which has a circulation of only 6,169. KNI also uses optical scanners to produce type automatically from a typewritten page; computer-controlled display screens for redesigning ad layouts; and a facsimile system for transmitting ad layouts. KNI's *Miami Herald* even became the first major metropolitan daily to paint its presses white and equip the pressroom with "demisters," making the area as clean as any other part of the plant. "What we did," says the younger Knight, "was turn the composing room into a manufacturing center."[1]

LONG-RANGE PLANNING ESSENTIAL

As a general rule, U.S. manufacturers of the past have had many enterprising and resourceful plant people who in most instances solved problems as they arose. Advancing technology and multiplying complexity in modern production systems have now made such old practices expensive and, largely, ineffective. And the relaxed approach of waiting for the competition to solve many critical problems first has become dangerous; and the equipment supplier today no longer has the prerogative to risk the considerably larger investments needed in time and money.[2]

Today, developments on the world industrial front call for a vigorous and imaginative program for progress in manufacturing automation. Are you planning to keep up?

Reducing costs by better taking advantage of labor-saving devices is basic industry practice. Much production management and engineering effort is spent in this activity. However, many productivity improvement opportunities are overlooked because they're evident only to the employees on the line who make and assemble the product or process the paper that is part of getting the product to the customer. These frequently untapped improvements are significant contributions of Reliance's cost reduction program.

Over the past six years, Reliance has been refining and extending a program which now delivers an average savings of 5 percent of the cost of sales. Among the most productive areas for cost reduction are:

1. Purchasing – reduced preprocessing by suppliers, bulk purchasing.
2. Manufacturing – improved material flow, more efficient tooling.
3. Administration – simplified forms, combined reports.
4. Product – reduced overdesign, simplified construction, standardized parts.

Each has come up for cost-saving scrutiny at many companies but with sproadic attention and sporadic results. Where Reliance has gained is in encouraging and acting on improvements at all levels of operations where daily activity can spell the difference between profit and loss.

The elements of Reliance's successful cost reduction program are aimed at detecting and counteracting program apathy. They include:

1. Total involvement and understanding.
2. Continuity.
3. Established goals.
4. Formal plan of action.
5. Assigned responsibility.
6. Integration with operations planning.
7. Measurement of results.
8. Peer group reporting.
9. Recognition of results.

While these are the basic elements, each division is free to tailor the program to its specific circumstances.

However, those knowledgeable in the real dollars and cents aspects can see the profit opportunities. Accomplishments of the highly successful minority in automation have been significant. Without a strong program your company is subject to real danger in that you may be tempted to expect your suppliers to carry the research and development burden. To lean on your suppliers excessively is to sap the creativity of your people and seriously limit your future profit-making capabilities.

Although it often looks like equipment builders are in an ideal position to develop new processes, *no one* is as close to the real problem or has the detailed knowledge of process problems and manufacturing costs as the manufacturer himself. Tackling the critical problem directly can have outstanding results. We are convinced that real success depends on constantly doing research on new methods and means of reducing manufacturing costs. Today, we are selling better products for prices little higher than those of five years ago — and in the face of increasing labor and material costs. Profit margins are as good today as then.

Clearly, as the need for automation grows, long-range manufacturing research and development becomes more and more imperative. The most important fact for management people to recognize is that useful manufacturing developments do not arise precisely as and when needed. Neither can they be expected to come without cost through the generosity of equipment makers. They must be *planned investments* in the future.

FORMAL PLANS

Without a plan, there is a great deal of "wheel-spinning." Each division prepares a manual which outlines how the program will operate. The basic elements of a manual are:

1. Corporate policy — the long-range cost reduction goals.
2. Division scope and objectives — specific areas for cost reduction and goals.
3. Procedures and programs — how the various programs involve workers, supervision, and division management.
4. Responsibilities — what individuals and committees supervise the program and their exact responsibilities.
5. Reporting forms — paperwork necessary to keep track of proposals and progress.

Keep the formal plan simple. Also, back it with adequate manpower to quickly process the cost reduction suggestions. Interest lags if employees see no action on their proposals. Success or failure depends largely on the people who are given the job of carrying out the cost reduction program. Don't assign responsibility too broadly or divide it up into too many pieces. There should be a single program coordinator who prepares the manual, expedites action, and audits the results.

Recognizing the need for and developing positive programs for cost reduction becomes a part of management's job.

CONCLUSIONS

In the future, good management practice will require that manufacturing and systems engineering teams be provided with the key tools for insuring profitable operations. A manufacturing research and development program will create the best bulwark against the competition ahead, *but* it requires a sympathetic understanding of the risks and difficulties involved.

Tomorrow's most profitable manufacturing innovations will result from today's manufacturing development work. Effective programs in force at present indicate that some amazing results can be obtained with an enterprising and resourceful program. In the process of creating the radically new approaches needed, there is no question that some blind alleys will be entered. But, fortunately or unfortunately, each of these are a part of the important learning process involved.

Manufacturing capability needed for a successful competition in the seventies unquestionably will call for a rising level of technical competence. More significant, however, will be top management's overriding influence stemming from sound basic policy on development and exploitation of automation.

FOOTNOTES

1 *Business Week*, August 29, 1970, p. 36.

2 *Outlook for Computer Process Control*, Bulletin 1658, U.S. Department of Labor, U.S. Government Printing Office, Washington, D.C.

REFERENCES

The Manufacturing Research Function by Harley H. Bixler, American Management Association Research Study 60, 1963.

Industry Case Examples

CASE 11A

Automated Tomato Harvesting

Latest tomato harvesting equipment separates ripe tomatoes by color — electronically (see Figure 1). Red, ripe tomatoes are channeled into the transport bins next to the harvester. Labor requirements drop from twenty eight to five workers. This research project is aimed at doubling the harvesting rate of process tomatoes.

CREDITS: Blackwelder Manufacturing Company, Rio Vista, California.

FIGURE 1.   Second generation automatic tomato harvester in operation.

CASE 11B

Research – Space Technology Applied to Cloth Cutting

The laser beam, whose introduction excited defense and space scientists a decade ago, today promises a development described by Genesco, Inc., the world's largest apparel company as "the first major advance in apparel manufacturing since the invention of the sewing machine."

The beam – often called "the light fantastic" – is shown at work in Figure 2 here in a computer-controlled machine, performing the basic operation of the business, the cutting of cloth to patterns for clothing. The demonstration was conducted by Genesco, Inc., in a factory of its subsidiary, L. Greif & Brothers; and by the system's developer, Hughes Aircraft Company of Culver City, California. Hughes' research laboratories achieved the first operation laser in 1960.

Lower Cost, Better Fit. Franklin M. Jarman, chairman of Genesco, concludes: "This introduction of space technology into the factory is significant not only for Genesco with its forty six operating companies, and for the rest of the apparel business, but for the overall economy. Consumers spend some $50 billion yearly for the industry's product – $18 billion for men's and boy's clothing and $32 billion for women's and children's." Jarmen cites as some of the benefits of this research:

1. Lower industry costs. Better value for and faster deliveries to the consumer.
2. Quick response to fashion changes, with better-fitting clothes.
3. Sharp reduction of large and risky inventories (now exceeding $3 billion).
4. A leg up on competition from foreign imports.
5. Cutting to a tolerance the width of a single thread with no errors and less waste.

This new system allows the cutting of one garment at a time with amazing speed and accuracy, rather than the standard dozen or more following the old method. The latter sounds impressive but actually handcuffs the industry.

How System Works. This automatic machinery shown in Figure 2 cutting men's suits can be adapted to cut material for virtually any articles of apparel, men's women's or children's. Four components largely constitute the laser cutting system: a computer storing programmed cutting instructions, a positioning device, the laser, and a conveyor. A single layer of material is unrolled from a bolt and moved along the conveyor until it is directly under the positioning device. Turned on by the computer, the laser's beam – but not the laser itself, which is stationary – is automatically directed and intricately maneuvered above the cloth following what can be a highly complex pattern stored on tape.

The beam cuts each garment according to programmed instructions that include directions to accommodate such matters as size and style. The conveyor then moves the cut material to where pieces are removed (see Figure 3), then another section of material moves into the cutting area.

Research. Hughes worked closely with the Genesco research department for two years in developing the optimum "one-high" cutting system and completing the prototype production machine shown. Genesco invested well over a million dollars.

Forty-two cutting methods were investigated, including blades, water-jets, thermal methods, chemical methods, and fracture methods such as high-velocity gas, before researchers settled on the laser. An important element in this research contrasts sharply with previous work which concentrated only on mechanizing the act of cutting cloth. This one was addressed to the problem of better response to the needs of the market. With this unit a man's sports jacket, a woman's skirt, and a child's pair of shorts, can be cut consecutively, without hesitation (See Figure 4). New styles, thus can be introduced at will.

The laser cutting system eliminates the necessity for accumulating large orders before starting up. Soon, if a customer orders just one, three or seven suits of a particular style, size, and material it will be possible at no extra cost to cut them almost solely by pushing a computer button. Whatever the retailer wants will be delivered in as little as half the usual time, and he can provide better service and fit to his customers.

CREDITS: Genesco, Inc.

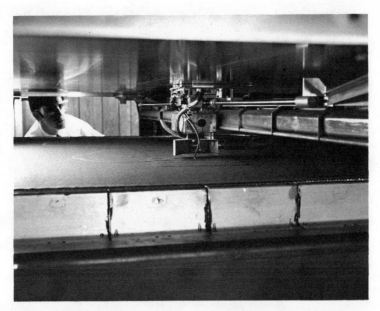

FIGURE 2.   Laser tailoring. The laser, a glamor development of the space age, is the heart of a new fabric cutter. The computerized device, shown here cutting a man's suit under the watchful eye of engineer Steve Toscano, was demonstrated for the first time at Fredricksburg, Virginia by Hughes Aircraft Company and Genesco, Inc.

FIGURE 3.   A laser leg. Bruce Campbell, Genesco technician, lifts the pant leg
of a man's suit just cut by the laser system.

FIGURE 4. Laser-cut suits. Engineer Steve Toscano (left) of Hughes and William Smith, Genesco research chief, watch as laser-cut suits roll from the laser system.

12

Plant Operating Policy

"It is the constant aim and tendency of every improvement in machinery to supersede human labor altogether." This is the comment of Dr. Andrew Ure, and it should be no surprise to anyone to learn that he wrote it in 1832. Most everybody who has seriously studied the technology of automation has learned that it is a new word for a process as old as industry itself — increase productivity and eliminate drudgery.

Industrial progress has never been the result merely of long-term accumulation of mechanization technology, skills, and production information. Students of industrial development conclude it has been influenced strongly by another technology as well — a social technology. It has occurred in stages, advancing by alternating steps — in one technology and then the other. It could be said that we are entering a third era of industrialization at present.

The early industrial technologies were apparent two hundred years ago when Adam Smith related the two traditional forms of the division of labor current in

227

advanced societies with the extension of its advantages by "those machines by which labor is so much facilitated and enlarged."

The modern industrial system was founded and grew from the knowledge of eighteenth century mechanical researchers, that natural events followed basic physical laws. In his effort to increase productivity, legend has it that Arkwright was first struck with the feasibility of mechanical spinning "by accidentally observing a hot piece of iron become elongated by passing between iron rollers." Thus, these craftsmen-inventors saw not just an interesting analogy but one process obeying a law which might also apply to a different and entirely new more productive process.

Along with Adam Smith's observation of the two technologies, a third step was being taken with the creation of the first successful factory by Strutt and Arkwright. By 1835 Andrew Ure discarded the basic principles of division of labor as outdated and misleading. He observed the industrial system as simply the factory system as developed by Arkwright — "the combined operation of many work people, adult and young, in tending with assiduous skill a system of productive machines continuously impelled by a central power."

TECHNOLOGICAL CHANGE AND INDUSTRIAL MANAGEMENT

Technological changes in the past have occurred through the birth and death of companies — still the prevalent way that achieves industrial change. Capital has usually been available for the exploitation of new devices or products. Any corporate structure created to exploit an invention was generally fairly rigid and identified primarily with the original application and product. For the most part new devices that rendered old ones obsolete have been exploited by new concerns. It was probably Elton Mayo who observed that, with the limited scale of business enterprise, change can take place without much real serious disruption in the social and economic order.

The growth in the numbers of industrial administrative officials, or managers, reflects the growth of organizational structures. Production department managers, sales managers, accountants, controllers, inspectors, training managers, publicity managers, research managers, etc., emerged to serve in highly specialized areas of general management as industrial concerns increased in size. These jobs were created by extensions or divisions of the original top manager's responsibilities over a long period of time. As a result it gave the whole social structure of which they are a part its present character.

Evolution of a rational bureaucratic system of control has made practical the increase in the scale of operations. To a singular extent, gradual separation of ownership and management has created the strong interest in survival of the enterprise and the existing management. More important, the growth of bureaucracy — the social technology which made possible the second stage of industrialization — was possible mainly because engineering technology has continued to advance in giant strides.

As a result, companies have grown in size; manufacturing processes have been routinized, mechanized, and advanced because coordination, planning, and monitoring could be subdivided into routines and taught as specialized managing activities.

Generally, industrial management proceeds on the basis of subdividing responsibilities into a practicable number of specialities and tasks. Each individual member then pursues his area or assigned task distinctly separate from the ultimate concern of the company as a whole. Top management is responsible for seeing to total relevance.

The technical requirements, duties, and responsibilities attached to each functional role are often precisely defined. Communication within management is largely vertical, i.e., between superiors and subordinates. Operations and activities are directed by instructions and decisions issued by superiors. This command hierarchy operates under the assumption that all goals and interests held by the company are, or should be, known only to top management. Management, often visualized as the complex hierarchy familiar in organization charts, operates simply with information flowing up through a succession of "filters," and decisions and instructions flowing downwards through a succession of "amplifiers."

The proper functioning of the company then depends on effective communications. This is, however, much more than a matter of providing information through a rigid paper notification of events and decisions affecting functionally related managers and department. It is also something more than providing for exchanges of information and opinion in meetings. What is essential is that nothing should inhibit individuals from applying to others for information and advice, or for exercising additional effort.

AVOIDING INFLEXIBILITY

To remain competitive in an age of automation, any plant must provide the utmost in flexibility. To avoid the "rigor mortis" of bureaucratic management, it is desirable to have few organization charts, and to define few functional responsibilities in minute detail.

An organic system of this kind, as contrasted with traditional bureaucracy, can be described as follows: Organic systems are adapted to unstable conditions, when problems and requirements for action arise which cannot be broken down and distributed among specialist roles within a clearly defined hierarchy.

Individuals have to perform their special tasks in the light of their knowledge of the tasks of the firm as a whole. Jobs lose much of their formal definition by interaction with others participating in a task. Interaction runs laterally as much as vertically. Communication between people of different ranks tends to resemble lateral consultation rather than vertical command.

DEVELOPING BROAD COMPETENCE

Managers, now and increasingly in the future, will have to develop more than mere technical competence. The importance of manufacturing systems knowledge cannot be overemphasized.

In too many cases industry is still concentrating on the competitive accumulation of premium research and development talent while underestimating its need of production systems professionals. Before the potential profit returns of automation can be realized, industry must concentrate in the foreseeable future on the overall needs of practical automation. Under the bureaucratic approach, these needs are all but lost in the pursuit of individual interests.

From the standpoint of practicality and cost, the manufacturing systems engineering approach for automation is a must. It is impractical to study and analyze methods *after* development of a complex processing system of any kind. The correct series of processing steps must be devised, planned and executed into equipment that is right the *first* time. Experience and know-how must be translated into a properly engineered system. Engineering the automated system must be a "before-the-fact" rather than the "after-the-fact" technique prevalent in the past.

Here, we can question whether the usual industrial engineering approach must be revised so as to focus activities henceforth on instigation of operations, rather than mere measurement of them. As far as automated facilities are concerned this is the case. Focus must swing to machine operations rather than manual operations. The ease with which human operators adapt to difficult processing operations is not readily duplicated in machinery. Concentration must be diverted from human motions. Success in automatic feeding, handling, transferring, positioning and like operations is dependent on inventive new thinking; simple duplication of manual operations is doomed to failure.

Today the manual functions defined by Gilbreth as Reach, Select, Grab, Return, Place, and Work are being replaced by mechanical units that Move, Hold, Orient, Position, Feed, Transfer, and Work. The ultimate goal is to feed raw material, hold it captive, and transfer from process to process through finishing, testing, and packaging for shipment. This is the real challenge.

THE LABOR PICTURE

Under automated production conditions, new problems arise in labor measurement, wage incentives, supervision, and cost control. Labor is no longer a key common denominator for productivity. Standard labor-hour costs can become unrealistic. An automatic production process involving no direct labor poses a real problem in costing out the product on a standard labor-hour system. Indirect costs and burden change in their importance. It becomes necessary to evaluate and develop a workable theory for machine-hour costs. Machine utilization takes on a new aspect. Some machines will be found to pay off under only part-time operation much like an automatic furnace that is called on to deliver heat only on demand of the system. In this regard, a whole new and highly sophisticated engineering philosophy is evolving.

Generally, what should management expect in these changing circumstances? Here, then, is the challenge. Automation is here to stay, indeed to grow significantly. In fulfilling the requirements for achieving profitable automation, there is a bright and promising future in tomorrow's industrial picture. But success does demand good communications.

How does any company explain the facts about automation to workers who find themselves replaced or displaced by automated production equipment? How do we explain it to the community? How do we communicate the prospects for future increase of automation (if any) to newly hired employees? How do we drive home the point that automation is a "must" to meet competition and stay in business? How does the company help find new opportunities for displaced workers, including opportunities for retraining and new jobs?

For the individual manager as well as company officers burdened with his individual what-to-do-about-it problem, all of the pertinent economic and background facts are helpful. They should serve as some grist for his communication mill. The fact that today's automation is only part of the past — and his own company's automation is only part of the total — demonstrates that this is a sweeping, urgent trend with dimensions in time as well as society.

But the next question must be: What can be done, what should be done? What program will assure full cooperation of all levels?

1. Communicate the economic facts of life to employees and the community. This should be general, i.e., pertaining to the economics of industry generally, and the nation. It should also be specific, i.e., what it takes to keep us in a position to meet competition.

Use display and exhibits: the competitors' products, presented in lobby or cafeteria, with details on what they sell for, and how much they cost to make.

2. Confide fully in all management and supervision. What's coming — what are men with blueprints and steel tape doing out on the factory floor? It is essential

that key people know what's in store — and why — so that they can help get across the facts with minimum misunderstanding when the time comes.

3. When — and, of course, if — it becomes necessary to lay off workers because of labor-displacing equipment, speak out frankly and fully. There's nothing to be gained by hoping the total number of jobs will be reported, by rumor and by guess, as smaller than they are. Rumor invariably makes them bigger. There's much to be gleaned by reporting the total as it truly is. *Be honest* and tell them if the competition has done a better job.

4. When — and again, if — new equipment comes upon the scene without causing loss of jobs, don't fail to speak up. Often, displaced workers may be immediately employed in other departments. While this plainly means fewer jobs for the community, necessarily, it also means no hardships to company individuals.

5. Spell out, in clear and ample wordage, just what could happen to the company if it neglected to keep up with competition.

6. Plan ahead; acquire more knowledge of the prospects for automation at all levels — executives, production, engineering, etc.

CONCLUSIONS

As industry moves gradually into more automated production systems, it is important that managers recognize the change in plant operating policy that invariably must accompany it. From the standpoint of practicality and cost, focus must change from manual operation to organized machine operations. The ease with which human operators adapt to difficult and changing process conditions is not readily duplicated in machinery. Quick shifts of personnel, addition or subtraction of operations, and fast hiring or layoff of workmen to match unexpected market changes become impractical and uneconomic.

Loss of the old simple piecework incentives for productivity call for a new look at employee morale builders based on much greater involvement in the entire process. Training for greater technical competence, increased supervisory skill, and more teamwork are needed. The entire plant staff group will develop through a massive evolutionary change to a considerably higher and more sophisticated level.

Accompanying these changes will come a far better, more realistic, and competently administered plant operating policy. For really effective results, no well-run plant will be without some form of employee relations program to explain and sell the benefits of the new plant operating policy. Thorough involvement and cooperation of the manufacturing team will be of tremendous value. Seldom will the companions in the manufacturing boat be found drilling holes in the bottom.

REFERENCES

Labour and Automation, Bulletin No. 1, The International Labour Office, Geneva, Switzerland, 1964.

Work, Workers, and Work Measurement by Adam Abruzzi, Columbia University Press, New York, 1956.

Progress Report – Automation Committee by Clark Kerr, Chairman, and Robben W. Fleming, Executive Director, United Packinghouse Food and Allied Workers AFL-CIO, Amalgamated Meat Cutters and Butcher Workmen of North America AFL-CIO, 1961.

Automation and Collective Bargaining by Benjamin S. Kirsch, Central Book Co., New York, 1964.

Automation – Economic Implications and Impact upon Collective Bargaining by John J. McNiff, Director, Department of Research and Education, International Brotherhood of Pulp, Sulphite and Paper Mills Workers, Second Edition, 1964.

Outlook for Computer Process Control, Bulletin 1658, U.S. Department of Labor, U.S. Government Printing Office, Washington, D.C.

Industry Case Examples

CASE 12A

Float Glass Computer System

A computer, closed-circuit television, and a video tape system (Figure 1) help Ford Motor Company manufacture "float" glass on a round-the-clock basis at the Dearborn (Mich.) Glass Plant.

The computer monitors about 500 "signals" — on conditions such as temperature, flow rate of liquids and gases, and levels in the glass-melting furnace and tin bath — at the rate of 30 signals a second.

The signals are transmitted to 200 control stations that regulate the glass manufacturing process in the melting furnace, the tin bath, the annealing furnace, and other float operating equipment.

A closed-circuit television network permits production operators to inspect visually the glass melting furnace and the ribbon of glass as it flows over the molten tin bath. The video tape system is used to record special manufacturing conditions.

The computer system provides instant information to production operators and also prepares management reports. It can be connected to data-processing equipment to further extend its uses.

The entire process control system is designed on the philosophy of "management by exception" — the computer alerts the operators only when some phase of the float process is abnormal.

At Ford Motor Company's Nashville (Tenn.) Glass Plant, three such float glass lines are in operation (see Figures 2 and 3). A single line is capable of producing a ribon of finished glass about ten miles long and eight feet wide every twenty four hours, and occupies more than 120,000 square feet of floor space and is more than 1,100 feet long. A ribbon of top quality glass is being produced within five hours after molten glass is drawn from the furnace.

As the glass ribbon emerges from the annealing furnace, it is inspected and automatically cut to pre-determined lengths. The alignment conveyor in Figure 4 guides the glass sheets through the cutters which trim the sheets into twin pieces to be laminated into safety-glass windshields or sizes for other glass sheet applications.

234

On a daily basis, the Nashville facility requires 7,000 horsepower to keep producing at planned rates and consumes 3½ million cubic feet of natural gas. Approximately one hundred hourly employees are engaged in running a single float line.

CREDITS: Ford Motor Company and Reliance Electric Company.

FIGURE 1. Electronic control of Ford Motor Company's new float glass facility in the Dearborn Glass Plant is maintained here in the plant's ultra-modern computer center.

FIGURE 2. A continuous ribbon of finished automative glass flows at the rate of 30 feet per minute from the new float facility.

FIGURE 3.   Control console for the float glass facility.

FIGURE 4.   Overall view of float glass system in operation.

CASE 12B

Automated Power Plant Coal Handling

Major considerations of plant operating policy are implicit in the continuous bulk handling system designed for the big new Cardinal power plant at Brilliant, Ohio. Whether coal arrives by barge or by train, the system can unload and stockpile it at a rate of 2,000 tons an hour.

To handle the 40 percent of the coal arriving by barge and the 40 percent by rail, this plant on the Ohio River has a continuous bucket barge unloader, and a rotary railroad car dumper (see Figure 5).

With 10,000 tons of coal consumed every day in its two steam units, maintaining the plant's desired 80-day reserve means heavy traffic in barges, trains, and trucks — and a big unloading job not easily done manually.

Barge Unloader. The advantages of the continuous bucket barge unloader becomes apparent: (1) because of its continuous operation, maximum speed or capacity are far higher; (2) only one man operates the unloader, from a pushbutton console; (3) maintenance is less — the bucket unloader is less subject to wear; (4) power consumption is lower, because the motors are designed for continuous digging and elevating of the coal, rather than for intermittent operation; (5) barge maintenance is also lower, because contact between buckets and barge is automatically controlled and doesn't depend on the skill or temperament of the operator; (6) capacity is unaffected by the water level; (7) poor visibility because of weather conditions doesn't stop unloading — the operator is located right over the barge.

How It Works. The barge unloader structure rises over 100 feet from offshore cells. The short-center digging elevator is mounted on a hoisting platform which lowers to barge level; the elevator also travels on a carriage which moves back and forth and sidewise. The system includes a barge haul, also controlled from the console.

The barge unloader is designed to handle barges 26 to 35 feet wide. It empties a barge in just two passes, with only a small amount of coal remaining. Unloading time for a 35-foot barge is about 67 minutes.

The digging elevator has buckets 10 feet wide which can scoop up nearly 1½ tons of coal every 2½ seconds as the barge is drawn slowly downstream. The 21 buckets are spaced at 4-foot intervals on massive steel bar-link chains which hang in a catenary loop below the foot sprocket.

On the first pass, the elevator is lowered into the barge, cutting deeply into the steep pile of coal. The barge moves along at a speed up to 10 f.p.m. Removing up to two-thirds of the coal on this pass, the elevator is then raised and the barge haul is reversed, quickly returning the barge to starting position for the second pass.

For this pass, the elevator is lowered — while digging — until the hanging buckets scrape the bottom of the barge hopper. Automatic traversing motion is started, and as the barge moves downstream, the elevator travels back and forth across the barge, removing the remainder of the coal.

The traversing cycle is synchronized with the downstream speed of the barge. The carriage has drives and tracks for both crosswise and lengthwise travel, so that an angular pattern is traced out, making a neat series of parallel cuts and cleaning up the coal evenly, right up to the sides of the barge.

The elevator discharges to a 54-inch-wide belt conveyor system, consisting of a horizontal transfer belt (Figure 6), on the hoisting platform, and a boom conveyor suspended from the platform and pivoted from a dock cell so that it can be raised and lowered with the digging elevator as the water level and barge draft change.

Rotary Car Dumper. The rotary railroad car dumper is designed basically to handle unit trains — dumping the cars one after another without uncoupling them, by rotating them on the axis of the car coupling. At present the plant is still receiving coal in conventional trains, and therefore each car is uncoupled before entering the dumper. Minor modifications will be required when the dumper begins to receive unit trains equipped with swivel-coupler cars.

The dumper clamps onto the cars, which weigh up to 300,000 pounds gross, and turns them upside down, emptying the coal into a collecting hopper. The dumping cycle is just 45 seconds from the time the car begins its 160-degree rotation until it is returned to its normal upright position. The entire unloading operation, including shifting of cars by means of a remote-controlled locomotive, is conducted by pushbutton (See Figure 7).

The dumper consists of two 32-foot diameter steel rings, turned on supporting rollers by means of 1-inch cables driven from a horizontal countershaft by a direct current adjustable voltage drive consisting of a DC Mill-type motor and a solid state conversion unit power supply. The coal car sits on a transfer table or platen with rails that line up with those outside the dumper.

To weigh each car, the dumper has a built-in electronic scale system, including a programmer, operator's control console, recorder, and calculator. The system records the name of the carrier, mine identification, car number, and gross, tare, and net weights of each car.

After the car is properly positioned and at rest in the dumper, and the selector switch placed on "automatic," the operator presses the "dump" pushbutton. Table locks disengage, permitting the scale to balance and record the loaded car's gross weight.

After a momentary delay, dumper rotation automatically begins at slow speed. Beam clamps descend simultaneously by gravity. At about 35 degrees, during which time the clamps have securely locked the car in place, the dumper goes into high speed until it is automatically stopped at about 160 degrees.

In about a second, the dumper automatically returns at high speed until it reaches the 35-degree point, where the speed slows so that the clamps can rise as it is braked to a stop in the upright position.

The scale balances and records the tare and net weights, then the table locks engage and the operator positions a new car in the dumper, ready for the next cycle.

CREDITS: Link-Belt Engineering Group, FMC Corporation.

FIGURE 5.   Coal barge unloader at work.

FIGURE 6.   Unloader conveying coal to storage.

FIGURE 7. Pushbutton coal car unloading system.

CASE 12C

Maintenance Monitoring

Few are the plants that have *not* made a very significant investment in a preventative maintenance program — a capable maintenance crew and a continuing technical training program for these people. Certainly all of the high-output, efficient, and profitable plants have approached this problem in one way or another. They must to satisfy management's goals of productivity from each machine by minimizing unscheduled downtime. The ultimate approach is the maintenance monitor which is capable of predicting failures "before they occur" and thereby permitting repair in a planned, orderly fashion, hopefully during the machines' next scheduled changeover. This would be typical of bearing failures in the machine, pumps, or drive system or other mechanical variations where vibration or temperature trends are indicative of equipment reliability.

Certainly it is "classic" to find the machine superintendent who compares his maintenance problems on a machine drive in this way: "With a mechanical-line shaft drive we can find the trouble in five minutes and it takes an hour to fix it — while with a sectional electric drive it takes an hour to find the trouble and then five minutes to fix it." Now picture the maintenance monitor that isolates drive problems and gives an *immediate* printed output as to which interlock or relay function opened first causing the shut-down or, in fact, which part of the drive system (AC power supply, regulator, driver, static power unit, or motor armature loop) is faulty and needs attention. This certainly minimized unscheduled downtime!

What Is It? The Maintenance Monitor is in fact a minicomputer, dedicated to monitoring machine vibrations, bearing and lubricant temperatures as well as drive system functions on a continuous basis (see Figure 8). The computer, once programmed and installed, does not have to be retrained or given vacations and time off for holidays. It merely stays on the job monitoring the machine process equipment twenty-four hours a day and giving an immediate alarm and tabulation of any unexpected variations.

One of the inherent advantages of the computer system is its speed. It can monitor and diagnose over one thousand digital signals in less time than it takes a standard control relay to pull in or drop out. Thus, the Maintenance Monitoring System can detect a process fault condition when it first occurs and instantly notify the operator. With suitable program routines, the computer can then continue its digital scan, and by correlating the existing conditions with those stored in its memory, actually diagnose the nature of the malfunction.

Functions. The modular design of the Maintenance Monitor allows it to be tailored to each individual application, and even expanded in the field. Some of the capabilities for monitored functions are as follows:

1. Process machinery vibration.
2. Motor vibration.
3. Bearing temperatures.
4. Process controller outputs.
5. Tachometer generator outputs.
6. Auxiliary interlocking and sequencing controls.

The system's flexibility also makes it easy to add other auxiliary equipment to the computer such as extra storage for process data and/or management reporting programs, and various output reporting devices such as CRT displays, Nixie tube displays and line printers.

Basic Drive Monitoring Functions. The basic maintenance monitoring system is designed to monitor drive system parameters by checking voltage and current limits, temperature signals, contact states, logic levels, synchronous alarms and preset or event counter outputs. This gives the system the capability of monitoring the following parameters:

1. Input power: drive and process.
2. Temperature: transformers, motors, and power modules.
3. Current balance: power modules.
4. Interlocking.
5. Sequencing.
6. Cooling-air flow: motors, cabinets and air ducts.
7. Regulation: major regulator loops and minor regulator loops.
8. Ground fault detection: control circuitry and motors.

Advantages. Although the initial cost of a maintenance monitoring system can typically range from $30,000 to $100,000, this cost can be paid off quickly by maintenance cost savings and increased production. By the same token, the initial investment is just that: a one-time investment that allows the plant to save money on maintenance training and equipment. In addition to providing a means of continuously measuring process performance and predicting impending breakdowns, the maintenance monitoring system also permits diagnosing problems and prescribing appropriate corrective action before an unexpected shutdown cripples production. Likewise, hardware and software are field-expandable, which means that the monitoring system can grow with the production facilities and need not become obsolete as process machinery and techniques change.

CREDITS: Reliance Electric Company.

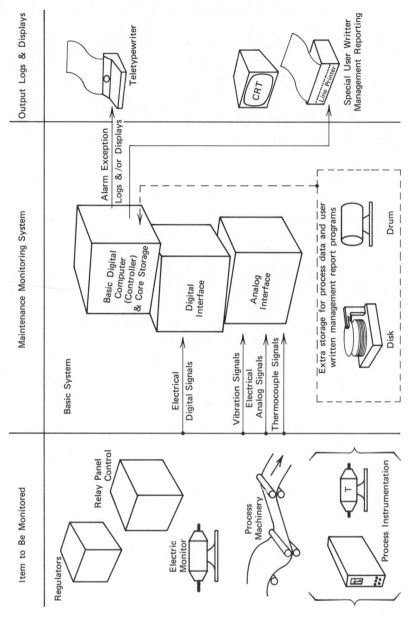

FIGURE 8. Functions of the maintenance monitoring system.

Major
Industrial
Change
Influencing
Automation
Planning

13

Effects on Industrial Makeup

Automation technology has created an anomaly. Although both laymen and industry people tend to see the *apparent* effects brought about by automation, the cause that in fact brought the automation into being is seldom visible. Not only does healthy business competition create the ever-present need to update facilities, so also does consumer demand. Today's powerful drive toward better safety, quality, and elimination of pollution has created the necessity for increasing automation to solve these problems.

To see the changing market needs and demands behind the growth of automation is important. The vital need to compete directly with low-priced small foreign automobiles is a case in point. Automation, as an aid to the men who man the plant, will provide the vital support for high-output, quality production where highly paid and highly skilled workmen are involved.

The ultimate result of the interworking of all these behind-the-scene influences is major industrial change. To compete profitably, whether the product is new or old, calls for careful assessment of this change in the light of any individual company's operations. Most critical from management's point of view is the fact that last minute recognition of the problem may spell disaster. To insure the profit potential from the changes growing out of automation of operations demands early recognition of the need. Time is of the essence.

Gradual and planned change to automation can resolve most of the critical aspects with only minor difficulties. A last minute "crash" program will either be doomed or more likely be costly and profitless.

PRODUCT MIX INFLUENCES

Automation can be planned for either of several directions — flexibility and change in products or for continuity and uniformity of long product runs. Either can be made profitable, but only by prior recognition of the requirements and the difference in automation planning.

Changes in product mix can direct planning into altogether different channels from those suitable where product character remains relatively stable. Here is where the usual management approach fails to solve the problem with any degree of success. The "tried and true" approach of shopping about for a piece of equipment to "automate our operations" simply cannot be used, for there is none.

Only through an innovative approach to practical manufacturing automation can an effective solution to truly flexible automation be obtained. There are, however, some serious barriers to effective implementation. In spite of the fact that we do have the technical capability to create profitable systems, industrial management in altogether too many instances either fails to comprehend the new rules for success or refuses to change from established traditional attitudes.

In the report of Panel 2 at the history-making Department of Defense-Industry Symposium on CAD/CAM/NC at Davenport, Iowa, October 1969, this basic problem of comprehension was placed in bold relief for managers.[1] The Committeemen comment:

A major issue and problem is the appreciation and acceptance by industry of the new manufacturing technology as a single system of production. It spans everything from the design of a new product to its distribution, including detailed design, specification, manufacturing engineering, materials management, fabrication of parts, assembly, test, warehousing, sales, and service. Numerical control involves the central elements of the sequence. Electronic data processing threads all the way through from end to end. Because of the breadth of this manufacturing system it vitally involves management, sales, finance, and personnel. Heretofore, industry tended to fragment and compartment itself and divide these functions into management subdivisions, each of which maintained its independence of authority and responsibility. Today it is still the wild exception rather than the rule for centralized numerical control operation coordinated across the board. Now the new technology inevitably ties together all of these functions and divisions and requires that they be treated as a system.

CHANGES IN MANAGEMENT CONCEPTS

Major industrial change, especially in the general management area as well as in the general organization of the plant, is in process as a result of market demands that make automation inevitable. One area of major change will involve services available. Changes in services will involve both the areas of services in plant equipment upkeep and maintenance and the services available to consumers.

Automated equipment brings with it not only increased complexity and sophisticated machinery, but the commensurate advance in the technical capabilities of the personnel who install, maintain surveillance over, and repair the elements that fail for whatever reason. Even though reliability has improved manyfold, the very great degree of added system complexity with automation can raise the maintenance spectre to critical proportions.

The direction of the corporate program is all important. Management concepts obviously must change considerably to avoid mere investigation without action. No shallow look-see, rule-of-thumb justifications, or partial answers will do. However, whatever action is inaugurated must also insure that the corporate program is practical and realistic.

CHANGES IN DIRECT/INDIRECT RATIOS

Obviously, one of the important considerations management must make in consummating the automation decision relates to its effect on the requirements for direct labor and indirect labor. In many cases, these ratios will change, and change radically. Chemical or metallurgical processing plants can run to personnel proportions as far out as one direct to fifty indirect, considering plant operation only. This is a radical change from the usual two to one ratio or greater in the everyday plant. Now these ratios are gradually changing in the same direction in metalworking plants.

Changeover to automated operations, therefore, should be approached on a well-planned basis. No overnight switch is to be recommended. In fact, those who plan well ahead, so as to introduce the change over a period of time find greatest success and greatest profitability. Even though ratios will inevitably change, the indirect service areas such as maintenance, engineering, and plant management become far more critical and far more important owing to greatly increased scope of knowledge and skills demanded.

This is especially true where plant or handling operations are computerized. This advanced level of control will offer no payback unless the personnel responsible for it can not only keep it in operation successfully but also continue to improve it in such a way as to insure close compliance with the ever-changing demands brought about by the competitive market.

Once the level of automation has been raised to the most satisfactory level possible, as it now is in some automotive parts suppliers' plants, the direct portion can actually fall to zero. All personnel may operate in a service capacity — setup, quality control, floor surveillance, materials handling, machine supply monitor, maintenance, etc. — with no actual direct labor being tied to a monotonous productive operation. It can be seen that successful change to this new situation calls for a complete switch from the old timeworn patterns.

It can be readily seen that the new skills and training, the new job demands, and the new management methods employed to maintain complex automated systems at full productive capability will be and are different. Much more influence will be present in the form of personal relationships that are conducive to a high degree of cooperation. Since better and more frequent communications are a critical must, this may be the key factor that makes computer information systems an additional requirement.

Without fast and correct action, the entire plant can easily shut down until the tieup is untangled. If a mandrel hangs up on a fast automatic tube piercing mill, the imperative is to know the operation is jammed and to take action to remove the tieup in minutes or risk shutdown of the entire upstream line for hours or days.

Thus, with the trend to automation the total of plant personnel may remain unchanged but what does change is their status into the higher skilled indirect category. Increased profit and increased wages result from the much better productivity obtained per man through skillful employment of equipment.

THE NEW EMPHASIS

Without question one of the key changes influencing automation planning for productivity is that of management attitude. Here, the new emphasis is subtle but nevertheless of striking importance.

The new emphasis is subtle since there still are few managers who evidence the knowledge that automation has in fact wrought a change in the necessary approach that they must employ. It is neither difficult to recognize nor is it difficult to miss completely.

That change is one that emphasizes the total solution to a manufacturing problem instead of looking for a quick solution in terms of available machinery. Every manufacturing system problem is unique. Seldom can an ideal solution be found that will put you in a one-up profit position without looking at your total system from raw materials to storage and distribution.

It is almost axiomatic that the automated systems providing maximum profit today were entered into by far-sighted management under the press of competition seen as a serious threat to future existence of the business.

SHIFTS AND CHANGES OF STATUS

One of the interesting results of this change in emphasis can be seen in the effects on the status of many companies. Once upon a time, most companies could be easily classified as to their status, capabilities, and their place in the roster of manufacturing. Today, this is much less the case. Automation technology is having a relentless influence on company character.

Although companies enter the field with a new single product required to create working automation lines, they seldom remain so.

A bit of observation will show that companies once devoted exclusively to construction of production machines are now creating much more of the total installation, including handling equipment and controls. Others once devoted to the materials handling business now handle the control package and associated equipment. Those who were only in controls now take on entire systems and may subcontract portions to other specialists. Then, the final development is the rise of the automation specialists who are devoted exclusively to the foregoing philosophy. They examine the problem, develop an economic strategy in terms of automation technology, engineer the needed equipment system to obtain the desired ends, build and deliver the working system. Implicit in the ultimate design is the use of a maximum of tried and proved elements – only the unique innovation of the system character is entirely new.

Thus, we see an entirely new and changing grouping of industries serving the automation needs of industry. Working closely with producing industry these new high technology groups serve the growing need to raise productivity and maintain profits.

FOOTNOTE

1 *Proceedings of the Department of Defense/Industry Symposium*, Davenport, Iowa, October 1969, U.S. Government Printing Office, Washington, D.C. p. 234.

REFERENCE

"Technical Innovation – Key to Manufacturing Success" by Roger W. Bolz, *Mechanical Engineering,* July 1969.

Industry Case Examples

CASE 13A

Computerized Controllers

A very significant advance in machine control is the so-called soft-wired systems. The soft-wired controller does everything that the conventional NC does, equally as well or better. In addition, it has a degree of flexibility far greater than its predecessors with their "hard-wired" logic. This flexibility can provide both a powerful alternative to DNC and, ultimately, a means to a more reliable DNC system. A "soft-wired" NC is used to control this 86-inch vertical turret lathe in the manufacture of turbine parts (see Figure 1).

Because the soft-wired controller uses a computer to do its logic it can be expanded by adding memory, input/output devices, and new or revised programs. It can also be conveniently data-linked to computers (see Figure 2).

Options can be supplied on the soft-wired control that permit editing of parts programs right on the floor. While editing may not be as "conversational" as with DNC, the speed of editing is still considerably greater than with conventional controls.

Parts programs can be stored inside the control simply by adding memory. This affords tapeless part production and avoids searching for a single program through an entire library.

An intermediate approach is to store frequently used patterns in the control's basic memory. The parts tapes can then call up these patterns with codes that indicate where they are to be machined and provides constants for variable dimensions. This pattern approach saves tape length and reduces the chances of programming errors.

Devices such as printers, tape punches, and readouts may be added to the soft-wired control so that it can provide production and maintenance data.

Because maintenance programs can be temporarily substituted for control programs, a high degree of self-checking can be attained with these controls. This can reduce considerably the analysis time associated with troubleshooting NC without adding additional failure-prone hardware.

Because of the memory expandability of the soft-wired control, it could be used in a DNC setup in which the central computer transfers partial or entire parts programs at high speed into the memory of the soft-wired NC. This

257

approach allows verification of the transmitted data prior to its use in controlling the machine. In addition, it reduces the duty cycle of the central computer, freeing it to direct more tools and perform other manufacturing functions (or allowing use of a smaller, less expensive central computer for the system).

Properly applied, DNC — or computer-aided manufacturing systems which include DNC functions — can revolutionize manufacturing and help overcome the cost-price squeeze faced by most manufacturers today.

When economic considerations do not allow the immediate installation of DNC, the soft-wired NC systems can provide an interim means to many DNC, savings. And if DNC is added later, plants already using the soft-wired controls can install better DNC systems at less cost.

Because of all these alternatives, the selection of the best system for any plant is not an easy one. However, the impact that the vast amount of NC technology available today can have on tomorrow's factory makes it essential that the manufacturer who hopes to maintain or improve his profitability make his plans right now.

CREDITS: N. L. Caban, Manager, NC Marketing and Sales, Industrial Systems Division, Westinghouse Electric Corporation, Buffalo, New York; "Today's NC and Tomorrow's Factory," presented at the 35th Annual Machine Tool Electrification Forum, sponsored by the Westinghouse Electric Corporation, Pittsburgh, May 25 and 26, 1971.

FIGURE 1. Soft wired N/C is used to control this 86-inch vertical turret lathe.

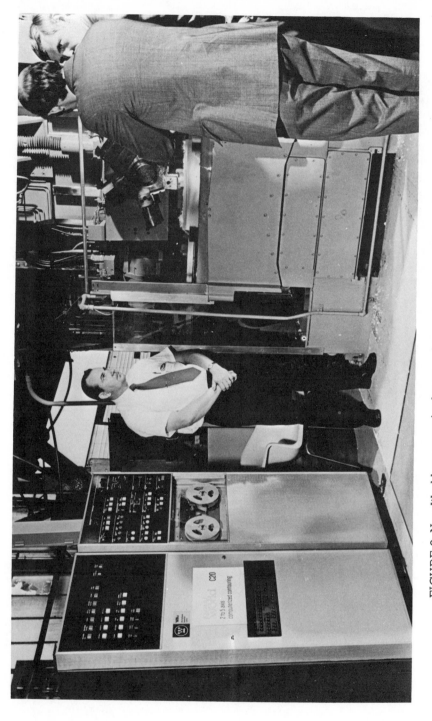

FIGURE 2. New World computerized contouring control for machine tools.

CASE 13B

Postal Service Streamlines Operation by Installing Computerized Key Cutter

In an effort to streamline the production of keys for postal lock boxes, the United States Postal Service has installed a single automated system that cuts, stamps, inspects, and packages them for mailing throughout the country. The system, controlled by a Digital Equipment Corporation PDP-8/S small computer, was designed and developed by Comstock & Wescott, Inc., a Cambridge, Massachusetts, consulting and engineering firm.

The Postal Service decided to automate this operation because of the continuing large demand for replacement of lost keys and for new keys brought about by the construction of new post office facilities and the transfer of lock box ownership. The system can process one key every 5½ seconds and one million per year based on a single 8-hour work shift.

Keys are ordered by coding a pre-addressed data card using a specialized portable device that punches holes in small numbered perforated boxes. Five of the boxes represent the cut code number on the postal lock box. A 5-digit cutting code number for this particular lock number, along with one for every other number conbination from 00000 to 99999, is stored in the computer's 250,000-word drum memory. These combinations represent a different pattern of notches for each key. The card also is coded with a single digit from one to seven, that specifies the number of keys ordered. These numbers are then read by a card reader in the system and the information processed by the PDP-8/S.

The manufacturing process begins at a special vibrator feeder that contains 1,000 blank keys. The feeder presents keys in the proper position to an inserting mechanism for movement at a loading station. The blanks are then cut by a radial group of five positioned cutters that are controlled by the PDP-8/S.

After the keys are cut, they are transferred to a milling station where a high speed contouring mill removes all heavy burrs and to a brushing station where the fine burrs are removed. The keys next move to a measuring station, where each of the five notches are checked for proper tolerances. The computer compares the depth of each cut with standards stored in its 4,096-word core memory.

Numbers that correspond to the cut code number are then stamped on the keys. The stamping process uses specialized stepping motors and DEC "W" and "R" series logic modules to link the stepping motors with the PDP-8/S.

Individual key orders are then brought to a packaging station where they are blister packaged on the original ordering card and dropped down a chute for immediate mailing.

CREDITS: Digital Equipment Corporation.

CASE 13C

Automation in Pipelines

Group 5A8 – Automation and Instrumentation of the American Petroleum Institute regularly meets to consider and study the needs for automation by the industry as well as the changing requirements and influences of the technology.

As a part of their activity, Group 5A8 has systematically surveyed the changing status of the industry for these purposes. As a result, a very clear picture emerges of the changing character of automation of pipeline operations.

The bar graph in Figure 3 was prepared from statistics drawn from this long-range study. It was felt that the percentage of unattended and remote controlled trunk-line stations (crude and products combined) was the most valid statistic for an objective measurement of automation. It is interesting to note that the percentage of stations with remote control exceeds the percentage of unattended stations in the current survey. This is perhaps a result of centralized data gathering and control of manned stations as the use of computers for system control become more common.

CREDITS: American Petroleum Institute Annual Report.

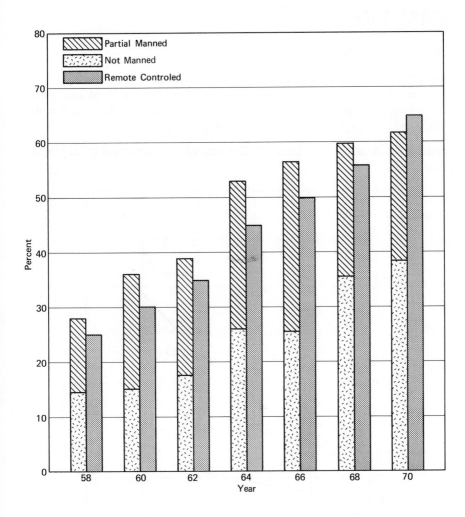

FIGURE 3. Trunk-line station automation.

CASE 13D

Changes in Services

From 1955 when services first made their appearance as a major development in the economy, growth has been continuous. In Figure 4 the trend since 1960 shows services outpacing both durables and nondurables from 1970 on into the future, creating a major change of notable proportions.

CREDITS: Based on data from U.S. Department of Commerce, the Conference Board.

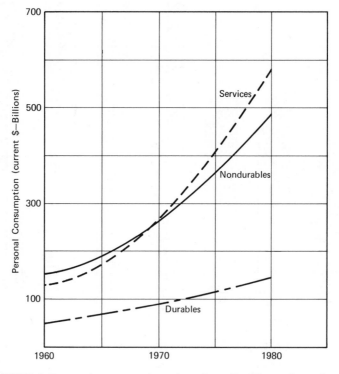

FIGURE 4. Personal consumption of services, durables and nondurables.

14

Effects on Productivity and the Enterprise

In a very real sense, the viability of an enterprise depends upon the total value of the goods or services that can be produced in terms of the number of employees required. Output per employee has long been accepted as a measure of productivity. Even though such a measure may not be universally acceptable, it does highlight one key fact: without automation, the practical and realistic output is directly limited by the capabilities of the manufacturing personnel (see Figure 1). Humans are limited with regard to performance on the production line. Given the most ideal of working circumstances, the manual worker can only be expected to produce what is physically practical. That which is beyond his physical capabilities simply cannot be done.

Given adequate tools with which to augment their capabilities, it is still a fact that physical limits are yet present to place severe restraints on individual productivity regardless of the number of workers employed.

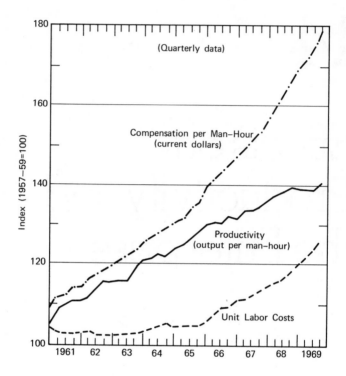

FIGURE 1. Productivity growth has not kept up with wage gains since 1965, so unit labor costs have moved up sharply. Unit labor costs represent compensation of persons per hourly unit of output. Data relate to total private sector. (Graph from U.S. Department of Labor.)

The products of the primary metals industries, those of the chemicals and petroleum industries, the pharmaceutical producers, paper-making industries, and a growing list of others are prime examples that dramatically illustrate the need for automated equipment not only to meet the productivity demands of this age, but as well to assure all those new and expanding requirements which elude any satisfactory solution by manual endeavor – pollution control, quality assurance, reliability, reasonable cost stability, product complexity, and a host of others. From 1959 to date, according to the National Industrial Conference Board, the petroleum industry has increased investment to $179 billion in new plant and equipment. This represents an increase in investment per employee of over 85 percent. Productivity is up 50 percent per employee.

THE PRODUCTIVITY PICTURE

Practically, then, productivity can be looked upon as the amount of goods and services produced by each individual. In simple terms it is really wealth. However, the National Industrial Conference Board[1] has studied this problem and, in order to apply further measures, has compared value added per employee, and value of shipments per employee, and value of shipments per production worker manhour.

These studies show that productivity of the first four companies in the top quarter of the same industry were at least half again as productive as all the others regardless of the measure used. Although there is some evidence that those companies with highest productivity held a greater share of the market, it is imperative that another conclusion be reached.

One can easily perceive that, based on these data, the high-productivity companies are the innovators – the early users of new technology and productivity-enhancing automation. When one makes a careful on-the-scene examination, it is in these companies where automation has first taken root and flourished. The laggards are mostly those who wait for new technology to be brought in at little or no research and development cost. Unfortunately, by the time this happens, the leaders are far ahead in a new and even more promising stage of advance.

RESEARCH AND INNOVATION

It is of utmost importance that the place of research and ultimately of technical innovation be kept closely in mind. In the future these aspects of the automation picture will become critical. Profit position will be directly related to the skill in developing and innovating. At present it is painfully obvious that the position generally in this area is not what it should be.

Dr. Philip H. Abelson, Editor of *Science* made this eminently clear in a recent editorial that strikes directly at the heart of the problem. His conclusion is that:

Our ability to compete in international trade is diminishing. In 1964 – a good year – U.S. exports exceeded imports by $7.1 billion. In contrast, during the first half of 1969 the value of exports topped that of imports by only $0.15 billion. An even greater factor than increasing imports of raw materials has been the invasion of foreign finished products such as steel and automobiles from countries that have more than recovered from the destruction of World War II. Our advantage of leadership in mass production techniques has disappeared. We still lead in scientific research and in the ability to innovate, but we have lost momentum[2].

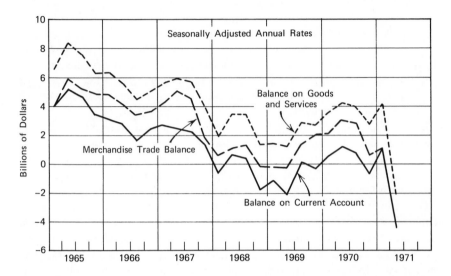

FIGURE 2. U.S. balances on goods, services, and transfers for 1971. Preliminary estimates for the second quarter indicate a merchandise trade deficit of $4.2 billion (seasonally adjusted annual rate) and a deficit on goods and services of $2.2 billion. (Graph from U. S. Department of Commerce, Council of Economic Advisers.)

Actually, the picture of international competition is dismal. The favorable balance of exports vs. imports has diminished precipitously to date – from the $7.1 billion figure to $2.1 in 1970, while preliminary estimates for 1971 indicate a total trade deficit of 4.2 billion. If all other cash flows are included the figure on all goods and services still shows a $2.2 billion deficit. (See Figure 2.)

The total lesson that must be learned and learned well is simply the one that if you wait until you *need* manufacturing automation, it is almost invariably because of a loss of business to competition, and then it is unquestionably too late!

Regardless of whether a company is large or small the problem is very similar. The product and manufacturing knowledge required may be lacking. To a larger and larger degree, the automated manufacturing system calls for a high level of intermixture of chemical, electrical, electronic, metallurgical, and mechanical technologies with an increasing interdependence of product attributes and manufacturing concepts.

Achievements of the required economically feasible system rely on ability to select and develop truly creative thinkers. Without such people there can be no research and development. Innovative ideas are delicate things that require favorable environment to grow. Management must establish receptive environment for this work. Policies, procedures, and facilities must be geared properly to the creative process. Such an environment opens the doors to new profits via several avenues: technically competent personnel capable of applying new scientific advances, the origin of new advanced manufacturing processes, technical skills that enable profitable "make" or "buy" decisions, cost reductions in overall operations, and increased product quality and/or reliability.

The results from such policies can be dramatic in terms of elimination of needless costs. The experience with a typical product under such a program is shown in Figure 3. Cost of the product was reduced to one-fourth that found with manual production.

However, the carefully organized program following the precepts outlined in the various chapters of this text take time, effort, and money. Devotion of management to these ends proves worthy of the goal. The chart shown in Figure 4 indicates in time scale how this cost reduction was achieved. Patience pays off.

A PRESENT OVERVIEW

In spite of the opportunities that are potentially available, one would be quite mistaken to conclude that automation is widely accepted throughout industry. On the contrary, even including the small "islands" or single small applications here and there, as previous comment has indicated, it is most probable that no more than 10 to 15 percent of our plants can be considered automated with extensive modern facilities.

These exist primarily in the highly competitive chemical, petroleum, consumer goods, foods, automotive suppliers, appliances, ordnance, electronics, and a few other industries. And, thus, there is good reason that this is so. Automation places a high premium on managerial planning. Experience has shown that it does not and should not happen by accident or for the mere sake of technical accomplishment. Necessity has mothered this technological advance.

Successful automation must be implemented where applicable as a result of an economic integration of product development, materials development, production development, and market development. For economic success there is no alternative. A growing industry must involve itself in manufacturing development or face the risk of technical obsolescence. Yet, even now there is a distinct lack of personnel trained to design, manufacture, install, control, and maintain highly automated production systems. Education to fulfill these needs will become more and more critical.

Myron Tribus has emphasized the lack of serious effort to keep up in many industries. His words should strike a resonant chord among many managers:

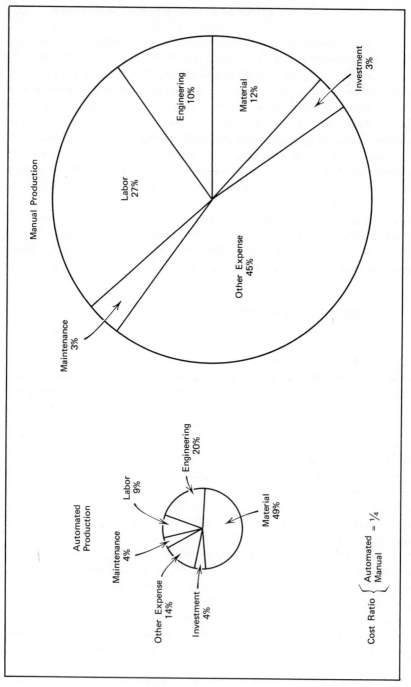

FIGURE 3. Cost comparison of similar components: automated vs. manual production operation at capacity level. (Courtesy Western Electric Co.)

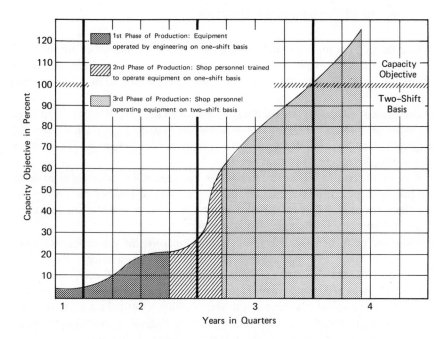

FIGURE 4. Production output of the product of Figure 3. (Courtesy Western Electric Co.)

After World War II this nation set about helping the rest of the world to recover and to master the latest technologies. We have been generous in exporting our technology. But at the same time we have not taken the steps to guarantee that our own civilian industries would keep up. We have been concentrating our efforts on defense, space, and atomic energy. In these areas, and in the technologies related to them, we have held commanding leads. But in the other areas of our economy, we are not in good shape. . . .

Technology is the essential ingredient in all of our profits — and it is the profit from our activities which supports our efforts for social advance. The funds for poverty programs, health programs, educational programs, urban-renewal programs, defense programs — all come from the economy. And if we falter technologically, the funds for these programs will begin to evaporate.

Technology helps in two ways: It can lower the cost of doing things. Whether we are producing goods or abating pollution, technology saves effort. It can increase the value of products by better design, greater reliability, greater service, greater aesthetic appeal, or use of substitute materials.[3]

SOME EDUCATIONAL NEEDS

Where do we look for personnel educated in the new technology of automation? Few schools as yet have awakened to the real need. At present, industries themselves must fill the void, but today their voices should be heard more often in academe.

Our present education system is largely geared to the past. It is poorly adapted to assist in meeting the needs of automation. To do so for the future, the emphasis must be redirected toward the creative, a talent which we are currently not developing to any significant degree.

Because of the obvious need and because this need will exert a singular impact on the development of automation, the Department of Defense has created the continuing Symposia[4] with industry leaders to develop realistic answers to all the related problems. They poignantly observe:

From the point of view of manufacturing managers, the present curricula in the junior and senior high schools, and even in the engineering colleges, do not provide the background, either generally or specifically, which is required to be effective in manufacturing. Too much of the current curricula is without meaningful application to reality. What is needed are more cogent courses in areas such as current manufacturing technology, social problems, creative writing, social development, and interpersonal relations.

Prerequisite to the discussion of educational implications is an understanding of the requirements of manufacturing organizations. No attempt will be made to detail what these requirements will be, but only to indicate in general what is now required in the way of education for factory employees and an indication of where this competence is now available and where it might be available in the future.

Their conclusion is to the point:

Most of the panels commented on the urgent need for effective educational and training effort to support the implementation of Computer Aided Design (CAD) and Computer Aided Manufacturing (CAM). The basic program must be one which supports the evolution, providing education and experience for our technological leaders now and in the future. To accomplish this, specific attention should be directed to the following:

1. Provide through the medium of trade association, professional societies, and government agencies a continuing program of seminars, short courses, and workshops.

2. Develop in conjunction with high schools, junior colleges and trade schools appropriate curricula supporting the basic prerequisites for technicians to effectively operate and service the machines and equipment involved in the CAD and CAM environment.

3. Establish formal engineering and manufacturing curricula leading to both BS and advanced degrees in CAD and CAM technologies.

Industry must actively participate in this effort. But, meanwhile, it will behoove managers to mount individual programs suited to their immediate needs. Productivity consciousness is not something that can be bought, compelled, or standardized. It will be necessary to create the atmosphere, the challenge, the motivation, and the wherewithal to achieve the benefits of true automation.

CONCLUSIONS

A year-to-year increase in productivity can be gained by continually planning improved methods. The way to step up the automation ladder is by studying new and different techniques and by making progressive improvements that will fit properly into operations in the future. A regular program of visiting noncompetitive manufacturing organizations will aid in this education. Active contact with local technical societies will upgrade performance and knowledge in a wide variety of individual fields of effort. In-plant seminars using both inside and outside experts should be a continuing management function. When it is recognized that relentless change will occur in manufacturing in the future, it immediately becomes evident that the development work for the methods to be used should be "in process" already.

One of the main avenues open for securing the great improvements in productivity required will be through step-by-step or progressive automation. Increased costs and foreign competition *can* be met with these innovative new methods and new equipment. The cost will be high, but, in turn, the positive results will be experienced over a long, long period of time.

FOOTNOTES

1 *The Conference Board Record*, July 1968, p. 13.

2 *Science*, September 12, 1969 vol. 165, No. 3898.

3 "Applying Science to Industry — Why America Falls Behind" by Myron Tribus, Senior Vice President, Xerox Corporation. Address presented at the Annual Meeting of the ASME, New York, December 2, 1970, "U.S. News & World Report," January 18, 1971.

4 *Proceedings of the Department of Defense/Industry Symposium*, Davenport, Iowa October 1961, U.S. Government Printing Office, Washington, D. C. p. 234.

REFERENCES

"The Big Questions in the History of American Technology" by George H. Daniels, *Technology and Culture*, vol. 11 no. 1, January 1970, p. 1.

"Mechanical Harvesting of Food" by Jordan H. Levin, *Science*, vol. 166, November 21, 1969, p. 968.

Productivity: A Bibliography, U. S. Department of Labor Bulletin No. 1514, July 1966. U. S. Government Printing Office, Superintendent of Documents, Washington, D. C.

Productivity: Key to Western Survival by Joseph Tod Meserow, Colonel U.S.A.F. Thesis No. 113, February 26, 1963, Industrial College of the Armed Forces, Washington, D. C.

"Technology and Manpower in Nonelectrical Machinery — the Decade of the 70s" by Lloyd T. O'Carroll, *Monthly Labor Review*, June 1971, p. 56.

"Is There Still Time . . . To Save U. S. Industry?" by Walter J. Campbell, Editor-in-chief, and Staff, *Industry Week*, October 4, 1971, p. S-1.

Industry Case Examples

CASE 14A

A Direct Numerical Control System in Operation

Over the past two decades, numerical control has provided many benefits such as higher productivity, shorter lead times and greater versatility in all types of manufacturing applications. There are problems, however, which have hampered even the best NC applications. Programming, tape punching, tape reading, storage of tape libraries and computer access rank among the top problems for large NC users.

During the planning of a new components plant at Winston-Salem, management of the Westinghouse Large Turbine Division realized that maximum utilization of the thirty NC machines being installed was a challenging and difficult goal. To meet this end, the Large Turbine Division – with the cooperation of the Company's Manufacturing Development Laboratory, Research Laboratories, and Computer and Instrumentation Division – developed a direct numerical control (DNC) system.

Among the basic objectives of the system are: (1) an increase in machine productivity, (2) an improvement in parts programmer efficiency, (3) a provision for management control, (4) adaptability to any machine tool/control combination, (5) high reliability, and (6) backup capabilities.

System Hierarchy. The system is composed of the three basic levels shown in Figure 5.

Level One is a large-scale time-shared computer, for instance a Univac 1108, a CDC 6400, or an IBM 360/75. For our system we have chosen the Univac 1108. The basic requirement for this computer is that it be capable of processing part programs written in a symbolic programming language such as CAMP or APT. Note that by using a remote computer connected to Level Two by a dial-up data link we have backup at Level One by placing a phone call to another computer.

Level Two is primarily communications terminal. Terminal equipment comprises a Univac 9200 with core memory of 16K (8-bit words), line printer, card reader, card punch, tape reader, tape punch, and a disc drive which provides an additional 3200K of on-line storage. Software has been developed to make the operation of this equipment self-supervising. Besides its use as a communications terminal, the 9200 provides many additional capabilities. One main feature

275

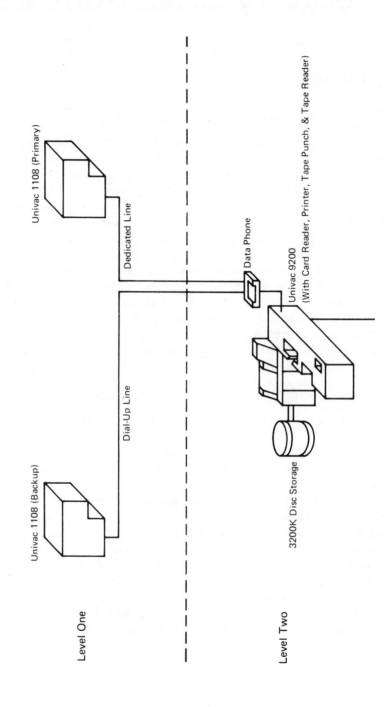

Univac 1108 (Primary)

Dedicated Line

Dial-Up Line

Univac 1108 (Backup)

Level One

Data Phone

Univac 9200
(With Card Reader, Printer, Tape Punch, & Tape Reader)

3200K Disc Storage

Level Two

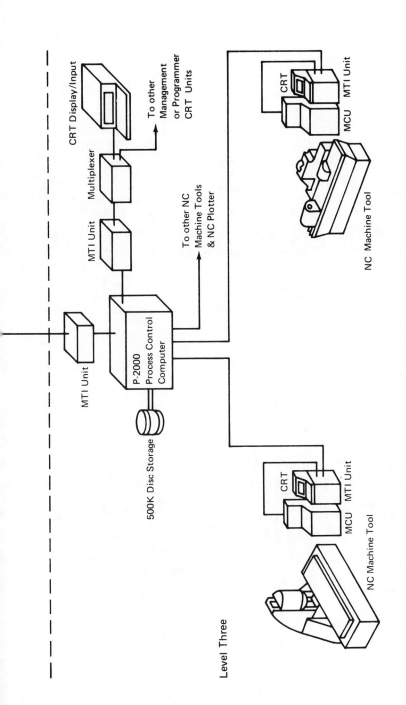

FIGURE 5. A direct N/C system for machine tools.

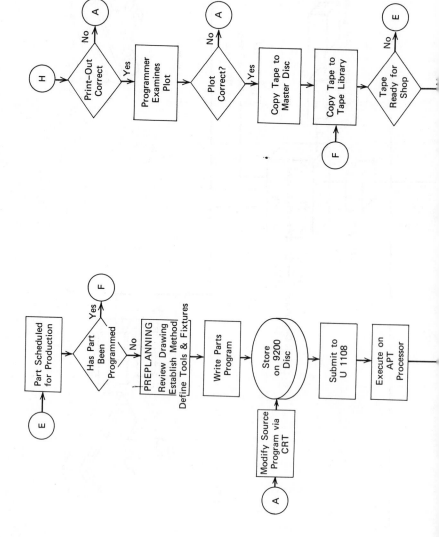

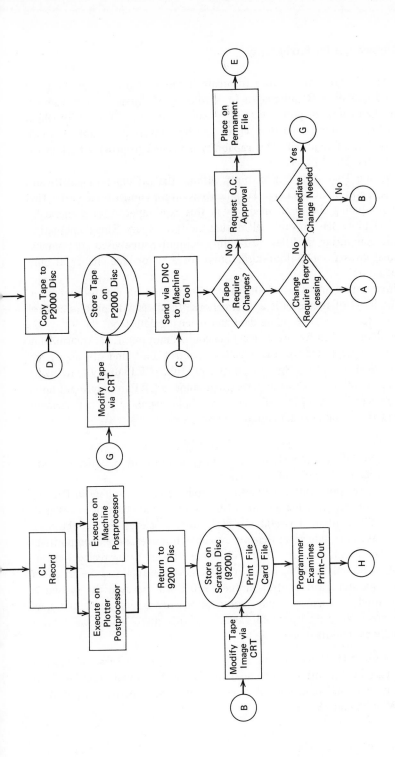

FIGURE 6. Flow diagram of DNC operations.

is the storage of data on its discs to create libraries of both symbolic parts programs and tape data. Requirements for hard copy of any data – for example, listings, punched cards, or punched tapes – can be met in any format for which a simple utility program is written. Cards can be converted to tape, and vice versa, in any format required. This capability is needed to provide backup for any failure on the Third Level.

Levels One and Two as described above, without the self-supervising software, have been in operation at the Large Turbine Division for almost three years. This method of producing the input media – in this case, tapes – for numerically controlled machines has proven itself during the three years. This combination of large-scale computing facilities used with a computer-supervised communications terminal gives floor-to-floor turnaround for a major tape change of only one to three hours.

Level Three represents the major change in numerical control operating systems. The heart of this level is a Westinghouse Prodac® 200 process control computer. Its tasks are (1) sending data and receiving data from the 9200, (2) storing tape data on disc for use as machine control unit input, (3) transmitting that data when requested to the machine control unit and its associated display/input device – in this case, a cathode-ray tube (CRT) unit, (4) receiving data from and sending data to the parts programmer's CRT display/input upon request, (5) receiving, storing, refining, and displaying data for management control, and (6) altering tape data upon request.

The flow of NC information, shown in the diagram in Figure 6, indicates many of the capabilities and benefits of the system.

The system began operating machines in a production atmosphere in January 1971. Since there are still some features of the system which have been designed but not yet implemented, it is still too early to feel or even predict the total impact of the system. Indicators that are available, however, have been positive.

Although the system is operational and its benefits are being realized at our turbine components plant, DNC is still in its infancy. The state of the art could be compared to numerical control in the 1950's. But because the system is software oriented, refinements and modifications are easily made and new applications are readily incorporated. Because of this, direct numerical control can continually expand its scope.

CREDITS: Robert N. Hallstrom and Howard T. Johnson, "A Direct Numerical Control System in Operation" presented at the 35th Annual Machine Tool Electrification Forum, sponsored by the Westinghouse Electric Corporation, Pittsburgh, May 25 and 26, 1971.

CASE 14B

Quality Control by Automation

Davis Gelatine, one of the world's largest manufacturers of gelatine, has reduced production costs and improved quality control in its newest and one of its largest extraction facilities — one that produces eight tons of dried gelatine each day — after placing it under the control of a PDP-8/L small computer.

Gelatine, which derives its principal value from coagulative and adhesive capacity, is made by boiling specially prepared skin, bones, and connective tissue of animals. The computer controls this boiling process by carefully regulating water temperature and its rate of flow into fourteen 2,500-gallon stainless steel containers, where gelatine is extracted from its raw materials. This process requires the control of numerous complex sequential operations similar to those found in the food processing and chemical industries. After final processing, the gelatine is cooled, dried, and packaged for distribution.

The computer and unrelated interfacing equipment are housed in a central control room shown in Figure 7, where a single operator views the entire manufacturing operation, via gauges, closed circuit television, and teletype-written data. The computer controls about 700 switch and detector contacts, 400 lamps and actuators, and set-points for about 60 analog channels. These monitor a large number of flow valves, pumps, and meters that interconnect production equipment in the plant.

The high-level control language developed for the system, designated BBCL1, enables plant engineers to write operational programs in a simple mnemonic language of micro-instructions resembling English-like statements. This language is learned in a few minutes and is capable of handling complex sequential operations from flowchart diagrams.

"Previously, this type of plant would have been controlled by stepping sequencers and hard-wired logic, but the computer offered many other advantages," noted T. N. Perrottet, a developer of the "turn-key" system by Bell Bryant, Ltd., New South Wales. He mentions "easy modification of the system's operating sequences; the availability of error printouts that alert the operator to any mechanical malfunctions or failures; increased reliability with the computer; and the possibility, because of the system's modularity, to easily control other parts of the plant by the computer."

CREDITS: Digital Equipment Corporation.

FIGURE 7. Tons of gelatine, for table desserts, photographic emulsions, and pharmaceutical tablets, are produced each day at Davis Gelatine's new extraction facility in Sydney, Australia. The entire process is monitored from this control room by a Digital Equipment Corporation PDP-8/L small computer located in the left background.

Index